AQA GCSE CHEMISTRY

Terry Mansfield, Ian Brandon, Rafael Fernandez

Second Edition

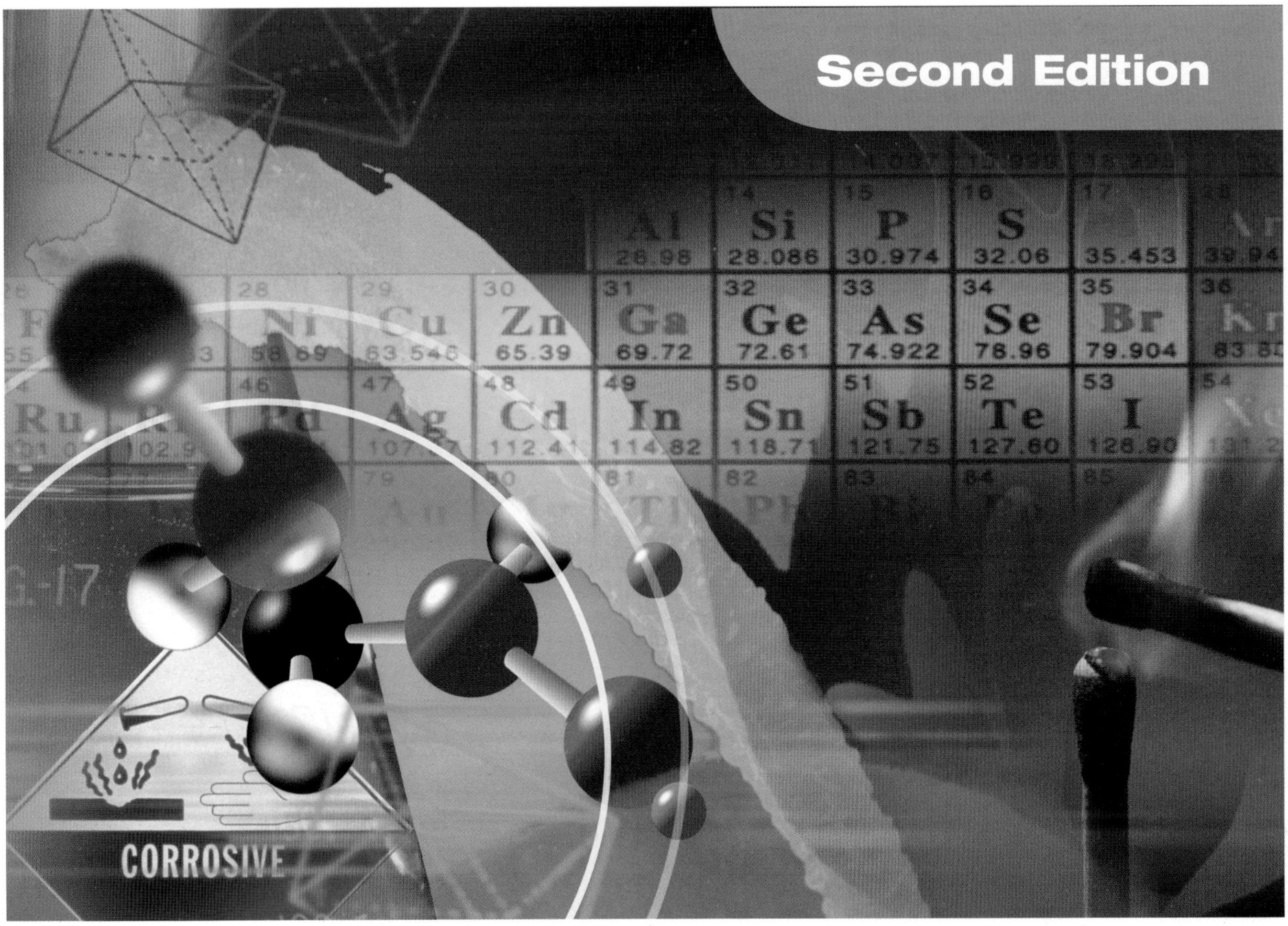

Hodder & Stoughton

A MEMBER OF THE HODDER HEADLINE GROUP

Photo acknowledgements

The publishers would like to thank the following individuals, institutions and companies for permission to reproduce photographs in this book. Every effort has been made to trace ownership of copyright. The publishers would be happy to make arrangements with any copyright holder whom it has not been possible to contact: Alcan Smelting & Power UK (58); Andrew Lambert (20, 123, 127 bottom, 96 bottom, 168 bottom, 175, 176, 177, 188 all, 189 all, 190); Corbis (44, 80, 119); GSF Picture Library (174 bottom left & right); Hodder and Stoughton (150 top left); Hulton Archive (2); Library and Reference Centre, Royal Society of Chemistry (169 left); Life File (127 top, 138, 151 all, 153, 158, 162 all, 163 top left & right); Martin Sookias Photography (147 all, 168 top left & right, 191 top); Poeton (Gloucester) Ltd (148); RD Battersby (134, 146 top, 163 bottom); The Royal Library, Copenhagen (169 right); Ruth Nossek (122); Science Photo Library (1, 16, 47, 63, 73 all, 79, 87 all, 90, 145, 149 top right & bottom, 150, 152, 157, 159, 160 all, 166, 173 bottom, 191 bottom left & right, 193); Stuart Wilson (172 all, 173 top, 174 top); Victoria and Albert Museum (64).

Orders: please contact Bookpoint Ltd, 130 Milton Park, Abingdon, Oxon OX14 4SB.
Telephone: (44) 01235 827720, Fax: (44) 01235 400454. Lines are open from 9.00–6.00, Monday to Saturday, with a 24 hour message answering service. Email address: orders@bookpoint.co.uk
You can also order through our website www.hodderheadline.co.uk

A catalogue record for this title is available from The British Library

ISBN 0 340 812 990

First published 2002
This edition published 2003
Impression number 10 9 8 7 6 5 4 3
Year 2009 2008 2007 2006 2005 2004

Cover illustration by Sarah Jones, Debut Art
Typeset by Fakenham Photosetting Ltd.
Printed in Italy for Hodder & Stoughton Educational, a division of Hodder Headline, 338 Euston Road, London NW1 3BH.

Contents

About this book

The contents

The contents of this book are designed to cover all aspects of the knowledge and understanding required by the AQA GCSE specifications in Chemistry (Co-ordinated) and Chemistry (Modular).

The subject content required by the KS4 Double Award specification for Materials and their properties attainment target is produced in a format identical to that used in the Hodder and Stoughton textbook *AQA GCSE Science*. This core material is supplemented by the additional subject content required for the specification in GCSE Chemistry.

What is in each chapter?

At the beginning of each chapter is a list of **key terms**. Where used for the first time, these key terms are emboldened. Some of the key terms are coloured. These are the extra terms you will need to know if you are going to be entered for the Higher tier papers in the final examination. All the key terms together with their meanings are also found in the **Glossary** on pages 195–202.

The contents of each chapter are divided into several **sections**. Each section concentrates on one topic. A symbol at the start of each section shows clearly which topic from the co-ordinated and modular courses is being targeted.

You will see a number of **Did you know?** boxes throughout each chapter. You will not have to learn the information in these boxes, but they are there to give extra interest to the topic.

At the end of various sections, you will find a number of **Topic Questions**. Because the topic questions have been designed to produce answers that you could use as a set of revision notes, it is recommended that you write down the questions as well as the answers. The questions written on a yellow background are the more demanding questions, expected to be answered if you are a grade B/A/A* student. Don't worry if you have to re-read the topic again when you try to answer these questions. This will help you to learn the work.

At the end of each chapter is a **Summary**. The summary provides a brief analysis of the important points covered in the section.

Completing each chapter are some **GCSE questions** taken from past AQA (SEG) or past AQA (NEAB) examination papers. The questions written on a yellow background are the more demanding questions expected to be answered if you are a grade B/A/A* student. Answering the GCSE questions will help give you an idea of what is wanted when you take your final science examination. Again, do not worry if you have to go back to read the work again. The examination questions may well test you on knowledge not included in the particular chapter. Don't worry – look through the other chapters to find the extra information you need to complete your answer.

Specification Matching Grid

Table 1.1 Materials and their Properties

Content			AQA specification references	
			Co-ordinated	Modular
Chapter	Section			
1 Getting together	1.1	Atoms	11.1/12.23	08/12 (10.1)
	1.2	Bonding	11.2	08 (10.2)
	1.3	Quanitative chemistry	11.8	07 (10.10)
2 Representing reactions	2.1	Representing chemical symbols, formulae and reactions	11.7	07/08 (10.9)
	2.2	Types of chemical reactions	Intro	07/08 (10.10)
	2.3	Exothermic and endothermic reactions	Intro	07/08 (10.10)
3 The Atmosphere	3.1	Changes to the Atmosphere	11.9	06 (10.11)
	3.2	Useful products from the air	11.6	07 (10.8)
4 The Earth	4.1	The rock record	11.10	06 (10.12)
	4.2	Useful products from metal ores	11.4	05 (10.5)
	4.3	Useful products from rocks	11.5	06 (10.6)
	4.4	Useful products from crude oil	11.3	06 (10.3)
5 Patterns of behaviour	5.1	The development of the periodic table	11.11	08 (10.13)
	5.2	Patterns in the periodic table	11.11	08 (10.13)
	5.3	Metals and the periodic table	11.11	08 (10.13)
	5.4	Patterns in the transition elements	11.12	05 (10.14)
	5.5	Patterns in the reactions of metal halides (halogens)	11.12	08 (10.14)
	5.6	Patterns in making metal compounds	11.12	05 (10.14)
6 Chemistry in action	6.1	Energy transfers in chemical reactions	11.16	07 (10.20)
	6.2	Reversible reactions	11.15	07 (10.19)
	6.3	Rates of reaction	11.13	07 (10.17)
	6.4	Reactions involving enzymes	11.14	07 (10.18)
7 Organic chemistry	7.1	The meaning of 'organic compounds'	10.4	21 (14.6)
	7.2	Burning organic compounds	10.4	21 (14.6)
	7.3	Homologous series	10.4	21 (14.7)
	7.4	Isomerism	10.4	21 (14.7)
	7.5	The physical properties of alkanes	10.4	21 (14.7)
	7.6	Alcohols	10.4	21 (14.8)
	7.7	Carboxylic acids	10.4	21 (14.9)
	7.8	Polymers	10.4	21 (14.6/14.9)

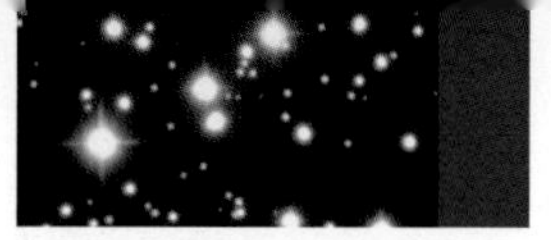

Chapter	Section		Co-ordinated	Modular
	Content		AQA specification references	
8 Industrial process	8.1	Sulphuric acid	10.7	22 (15.1)
	8.2	Aluminium	10.7	22 (15.2)
	8.3	Titanium	10.7	22 (15.3)
	8.4	Steel	10.7	22 (15.2)
9 Aqueous chemistry	9.1	Water is essential to life	10.15	21 (14.1)
	9.2	The water cycle	10.15	21 (14.1)
	9.3	Hard and soft water	10.15	21 (14.1)
	9.4	Solubility	10.15	21 (14.2)
	9.5	Acids and bases	10.15	21 (14.3)
	9.6	Making salts	10.15	21 (14.4)
	9.7	Measuring the concentrations of solutions	10.15	21 (14.5)
10 Detection and identification	10.1	Laboratory methods	10.16	22 (15.4)
	10.2	Instrumental methods	10.16	22 (15.5)

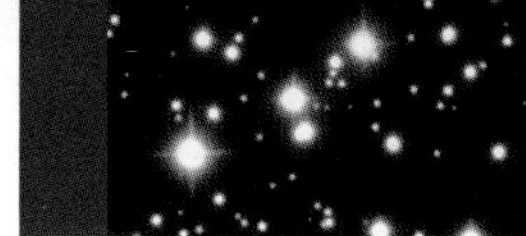

Ideas and evidence in Science

You will find that many sections contain information which is marked with a bell and a vertical stripe in the margin. This is material to support the 'Ideas and Evidence in Science' part of your course. It will provide you with information about:

- how scientific ideas were developed and presented,
- how scientific arguments can arise from different ways of interpreting the evidence,
- ways in which scientific ideas may be affected by the contexts in which it takes place (for example, social, historical, moral and spiritual) and how these contexts may affect whether or not ideas are accepted,
- the problems science has in dealing with industrial, social and environmental questions, including the kinds of questions science can and cannot answer, uncertainties in scientific knowledge, and the ethical issues involved.

Each of the 'Ideas and evidence' contexts needed for whatever course you are following is included in this book. A guide to these contexts and whether they are needed for Core or Higher tier is given in Table 1.2.

Table 1.2 Contexts for the delivery of 'Ideas and evidence' in Materials and their Properties

Section	DA	Core/ HT	Context
1.1	✓	core	How the idea of the atom became generally accepted after Dalton reintroduced the idea about 200 years ago
3.2	✓	HT	How economic factors affect the conditions under which the Haber process is carried out
3.3	✓	core	How benefits from the use of nitrogenous fertilisers need to be balanced with the potential contamination of water supplies
4.4	✓	core	How the burning of hydrocarbon fuels affects the environment
4.4	✓	core	How the disposal of plastics affect the environment
5.1	✓	core	How early attempts to classify elements systematically led to the development of the modern periodic table
5.1	✓	core	Why the periodic table gradually became accepted as an important summary of the structure of atoms
6.4	✓	core	How the use of microbes and enzymes to bring about chemical reactions has advantages and disadvantages
7.2	✓	core	Comparing the cost, efficiency and cleanliness of burning different fossil fuels
9.5	✓	core	How the contributions of Arrhenius, Lowry and Brønsted affected our understanding of acid-base behaviour.
9.5	✓	core	Why the work of Arrhenius took longer to be accepted than that of Lowry and Brønsted

Some hints about doing well in the final written examinations

Some frequently used command words and what they mean

Before you can answer a question, you need to know what is expected. Question-writers use command words or phrases that inform you of the style of answer they expect you to give. A list of the most frequently used command words and phrases is given below. Question-writers assume that you have learned the meanings of the words or phrases.

Command word	Meaning
Calculate or **work out**	means that a calculation is needed together with a numerical answer.
Compare	means that a description is needed of the similarities and/or differences in the information that has been provided.
Complete	means that spaces in a diagram, a table or a sentence or sentences need to be completed.
Describe	means that the important points about the particular topic must be provided.
Draw a bar chart	– if the axes are already labelled and scales have been given then the values given must be plotted as bars. – if the axes are labelled but no scales have been given then scales need to be added and the values given need to be plotted as bars.
Draw a graph	– if the axes are already labelled and scales have been given then the values given need to be plotted as points and a line (or curve) appropriate to the points plotted must be drawn. – if the axes are labelled but no scales have been given then the scales need to be added, the values given must be plotted as points and a line (or curve) appropriate to the points plotted must be drawn.
Explain how or **Explain why**	means that scientific theory must be used to show an understanding of how or why something happens.
Give a reason or **How** or **Why** ...	means that the answer requires a cause for something happening based on scientific theory.
Give or **Name** or **State** or **Write down**	means that a short answer with no supporting scientific theory is needed.
List	means that a number of short answers are needed, each one being written one after the other.
Predict or **Suggest**	means that the answer is based on a *consideration* of various pieces of information and suggesting, without supporting theory, what is likely to happen.
Sketch a graph	means that a line (or curve) is to be drawn to show a trend or pattern without the need to plot a series of points.

Use the information	means that the answer must be based on the information provided in the question.
Use your understanding of ... **to**	this is the science topic around which the answer needs to be built.
What is meant by	means that the answer is likely to be a definition.

Some more hints

Obviously if you want to do well you need to have learned and understood as much as you can. However here are some hints about answering questions.

- Do not rush – no marks are awarded for finishing first. A paper worth 100 marks is designed to allow you about 90 minutes to finish it. This means that you have nearly one minute of time to think and write down 1-marks worth of answer.
- Read each question carefully at least twice before you write down your answer. If you need to do rough working to sort out your thoughts use the gaps in the margins – but make sure you put a line through this rough working.
- Look at the number of marks awarded for each part of each question. Each mark is given for a different piece of information:
  - *1 mark* means that one piece of information is needed.
  - *2 marks* mean that two pieces of information are needed etc.
- Lots of questions ask you to give a reason for something or to explain something. Such answers are usually worth 2 or more marks. Your answers to these should include a 'because' part.
- Do not throw away marks. Marks are often given for:
  - units such as joules, °C, ohms etc. Learn all the units and what they measure.
  - the names and symbols of chemical elements – so learn them
  - equations, such as 'potential difference = current × resistance' – so learn the equations and how to use them in calculations. Remember that all equations need an '=' sign in them.
- If you are writing an answer that needs several sentences, make sure that each sentence is saying something new and is not just rewording an earlier sentence.
- Try to avoid using the words 'it', 'they' or 'them' in an answer. The marker may find it difficult to understand what you mean.
- Take care when you are drawing graphs. Make sure all the points are correctly plotted. When you draw in the line for your points use a pencil with a fine point and try to draw the complete line in one go.
- If you have learned the work you should finish the paper in good time. Go through the paper again and check what you have written – it could save you throwing away some marks for silly mistakes.

Chapter 1
Getting together

Key terms

anions • atoms • atomic number • cations • charge • compound • covalent bond • electrolysis • electron • element • formula mass • free electrons • ion • ionic bond • isotope • Law of Conservation of Mass • mass number • molar volume • mole • molecule • negative • neutron • noble gas • nucleon • nucleus • Periodic Table • positive • proton • relative atomic mass • relative formula mass • relative molecular mass

1.1 Atoms

Co-ordinated	Modular
11.1/12.23	08/12 (10.1)

Although the information provided in this chapter is based largely on work carried out on the theory of atomic structure developed in the last 80 years it is important to be aware of the historical background that led up to the modern ideas of atomic structure.

The story of the atom – from the Greeks to John Dalton

The idea that all substances could be made of atoms originated in Greece some 2000 years ago. At that time many Greek thinkers (philosophers) were interested in trying to determine what all substances were made from.

Figure 1.1 *Democritus*

Democritus (460–370 BC) was a philosopher who believed that if a lump of metal was cut into smaller and smaller pieces you would eventually end up with a very small piece that was too small to cut up further. He called these very small pieces 'atomos' (which means indivisible). He believed that all atomos were made of the same matter but were different in shape, size and speed.

However, another Greek philosopher called Aristotle (384–322 BC) believed that it was possible to keep dividing up something into smaller and smaller pieces. He did not believe in the idea of atoms. He believed that all substances were made up of four 'elements' – earth, air, fire and water and that each substance was made of a mixture of the four elements.

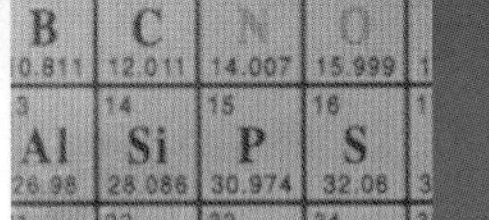

Democritus's idea was not accepted because it required people to believe in the concept of invisible particles whose behaviour could be explained and was therefore not influenced by the gods. Aristotle's ideas gained favour because they were easier to understand. They were accepted for nearly 2000 years.

Did you know?

For many centuries the ideas of Aristotle helped:

- Egyptian and Chinese philosophers and metalworkers to investigate ways to turn common metals into gold. The ability to turn common metals into gold was known as transmutation – later called alchemy. The Egyptians wanted the gold because of its monetary value. The Chinese wanted it because it was thought to give long, if not everlasting life for those who swallowed it.
- Arabic alchemists attempt to discover the 'philosopher's stone' which would not only help transmutation but would provide both wealth and health.

No-one succeeded in turning common metals into gold or in finding the philosopher's stone but a number of important chemical processes were discovered, a knowledge of practical chemistry was gained, as well as a better understanding of chemical reactions.

All these advancements in knowledge and experience began to undermine the theory of Aristotle.

In the 18th and 19th centuries chemists began to realise that many solids, liquids and gases could be broken down into simpler substances called elements. They also realised that elements could be joined to make more complicated substances called compounds.

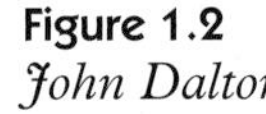

Figure 1.2
John Dalton

Early in the 19th century John Dalton (1766–1844), a British chemist and physicist, developed his atomic theory which attempted to explain the structure of elements and compounds. This theory was based on the concept of atoms proposed by Democritus almost 2000 years before.

Dalton's atomic theory was made public in a series of lectures and in a book called *New System of Chemical Philosophy* which was published between 1808 and 1810.

The key parts of his theory proposed that:

- all elements are made up of very small solid spheres called atoms
- the atoms of a given element are identical and have the same weight
- the atoms of different elements have different weights
- chemical compounds are formed when atoms combine
- a given compound is always made up of the same number and type of atoms.

There were some important scientists, such as Humphrey Davy and Michael Faraday, who were still reluctant to accept the idea that all matter was made up of atoms. It was not until the late 1800s that the existence of atoms was finally accepted. By this time Dalton's ideas had been successfully used to explain the existence and structure of molecules and in the determination of the relative atomic masses of the elements.

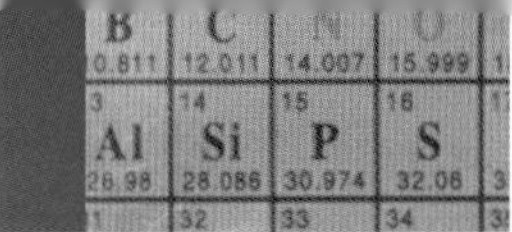

Through the work of John Dalton it is now known that the basic building blocks of all materials, whether they are solids, liquids or gases, are tiny particles known as **atoms**. Atoms rarely occur on their own – usually they join up with other atoms. There are about 100 different sorts of atoms.

If all the atoms in a material are identical, the material is an **element**, but if atoms of two or more different types are chemically joined together, the material is a **compound**.

Atoms are the smallest particles of an element that can exist on their own. Dalton believed that they could not be broken into anything simpler. Research carried out during the last 100 years has shown, however, that atoms consist of even smaller particles. The most important of these are called **protons**, **neutrons** and **electrons**.

The structure of the atom

The atom is now known to consist of a very small **nucleus** containing protons which have a positive **charge** and neutrons which are uncharged. Negatively-charged electrons orbit the nucleus in shells or energy levels. The negatively-charged electrons are attracted to the positively-charged nucleus.

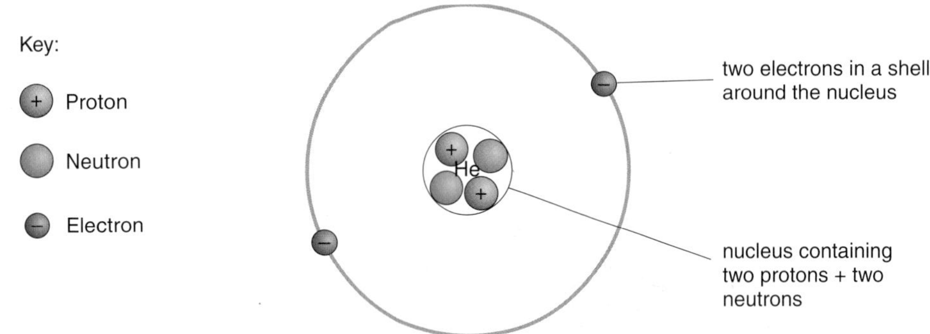

Figure 1.3
An atom of helium

Atomic number (Z)

Each element in the **periodic table** has a fixed number of protons in its atoms. For example, all hydrogen atoms have just one proton in the nucleus, all helium atoms have two, all lithium atoms have three, etc. This number of protons is known as the **atomic number** (proton number), Z. Since there are always equal numbers of protons and electrons in any atom, the atomic number also tells us the number of electrons present in the shells around the nucleus.

This atomic number is usually shown as a subscript to the left of the symbol used for the element e.g. $_1H$ or $_2He$ or $_3Li$ etc.

More about the particles

Atoms have no electrical charge because the **positive** charges of the protons are cancelled out by the **negative** charges of the electrons. In all atoms of a particular element there are equal numbers of protons and electrons.

Most of the mass of the atoms is due to the protons and neutrons in the nucleus; electrons have very little mass as Figure 1.4 shows.

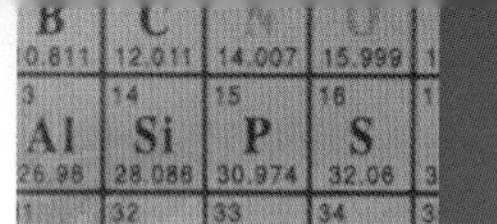

Figure 1.4

The relative masses and charges of the particles in atoms				
Name of particle	**Symbol**	**Relative mass**	**Relative charge**	**Where found**
proton	p	1	+1	nucleus
neutron	n	1	0	nucleus
electron	e	negligible	−1	in shells

Did you know?

If an atom was the size of Wembley Stadium, the nucleus would be smaller than the spot at the middle of the centre circle where the ball is placed for the kick off!

The radius of an atom is only about 10^{-8} cm (0.000 000 01 cm) and the radius of a nucleus is about 10^{-12} cm (0.000 000 000 001 cm).

Mass number (A)

To get some idea of the relative mass of different elements it is necessary to take into account the fact that most of the mass of the atom is due to protons and neutrons. The **mass number** (nucleon number), A, is obtained by adding together the number of protons and neutrons (**nucleons**) present in the nucleus. So in a helium atom, where there are two protons and two neutrons, the mass number is (2 + 2) = 4.

This mass number is usually shown as a superscript to the left of the symbol used for the element e.g. ^{4}He.

The atomic number and mass number of an element can both be displayed together, so an atom of sodium (Na) that has 11 protons and 12 neutrons in its nucleus would be shown as $^{23}_{11}$Na (Z = 11; A = 11 + 12 = 23).

Isotopes

Although the number of protons in the nucleus of an atom is fixed for that element, the number of neutrons present in the nuclei of atoms of the same element can vary from atom to atom. Atoms that have the same number of protons but different numbers of neutrons in their nuclei are called **isotopes** and most elements exist as mixtures of isotopes. An example of this is lithium which has two isotopes: $^{6}_{3}$Li and $^{7}_{3}$Li.

^{6}Li has three protons and three neutrons but ^{7}Li has three protons and four neutrons, so ^{7}Li is slightly heavier than ^{6}Li. Naturally-occurring lithium is a mixture of about 90% ^{7}Li and 10% ^{6}Li giving an average relative mass for a lithium atom as 6.9. This average relative mass which takes into account the different masses of the isotopes and their relative amounts in a naturally-occurring sample of the element is called the **relative atomic mass**. Usually the relative atomic mass is rounded to the nearest whole number.

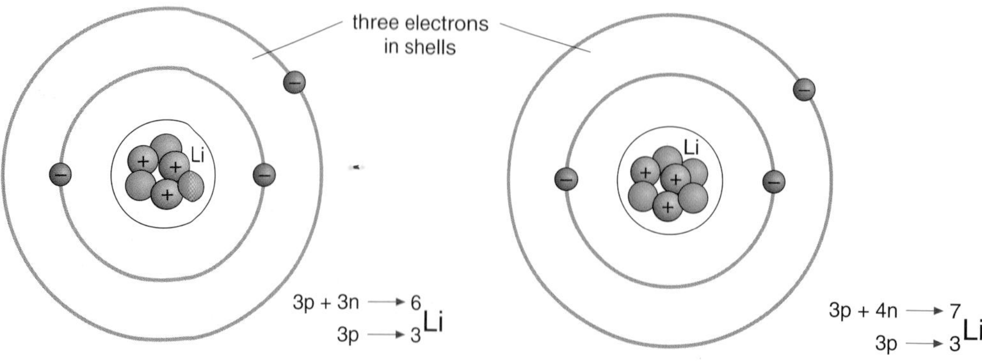

Figure 1.5
Isotopes of lithium – atomic number 3

Remember that during chemical reactions the number of protons and neutrons in the nucleus *never changes*.

Arrangement of electrons

Each electron in an atom is at a particular energy level (shell). The electrons in any atom occupy the lowest available energy levels (the innermost available shells).

The simplest atom is an atom of hydrogen which has one proton in the nucleus and one electron at the lowest energy level (in the first shell) outside the nucleus (see Figure 1.6).

The second element is helium. It has an atomic number of 2 so it will have two electrons outside the nucleus. Both of these electrons go into the first shell (Figure 1.7).

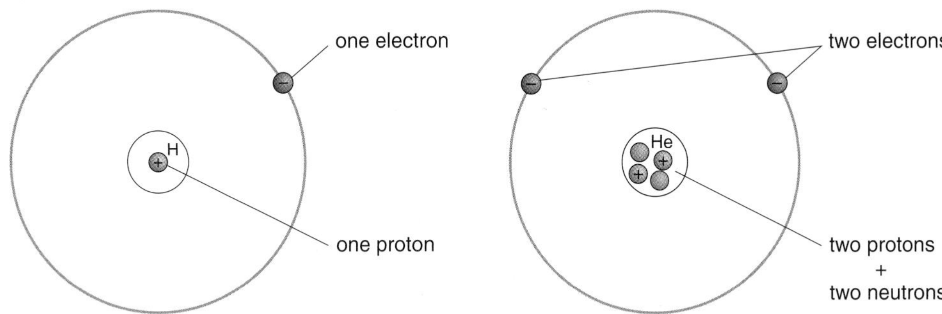

Figure 1.6 *The atomic structure of hydrogen (no neutrons present)*

Figure 1.7 *The atomic structure of helium*

With two electrons in it, the first shell is full. The evidence for this is that helium is a **noble gas**. All noble gases exist as individual atoms and do not take part in chemical reactions because their electronic structures are very stable. If an atom of an element has more than two electrons, the extra electrons go into the second shell and this is what happens with element number 3 – lithium.

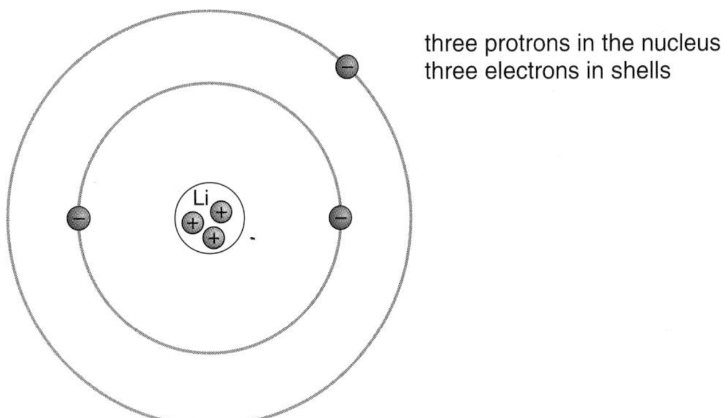

Figure 1.8 *The atomic structure of lithium (neutrons not shown)*

As the atomic number continues to increase and the number of electrons also increases, the second shell begins to fill up (see Figure 1.9).

In element neon, atomic number 10, the second shell is full – neon is a noble gas. The second shell is stable and full when it contains eight electrons. Any further electrons go into the third shell until it too becomes 'full'. This happens with element number 18 – argon. Argon is also a noble gas so its outer shell must be full when it contains eight electrons. In elements 19 (potassium) and 20 (calcium) the fourth shell is used.

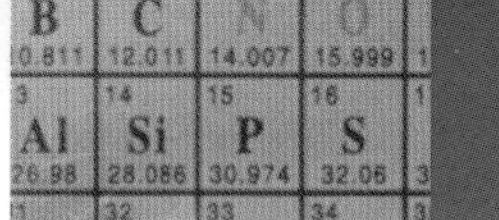

Figure 1.9
The electronic structures of elements 3 to 10. No details of the contents of the nucleus are shown

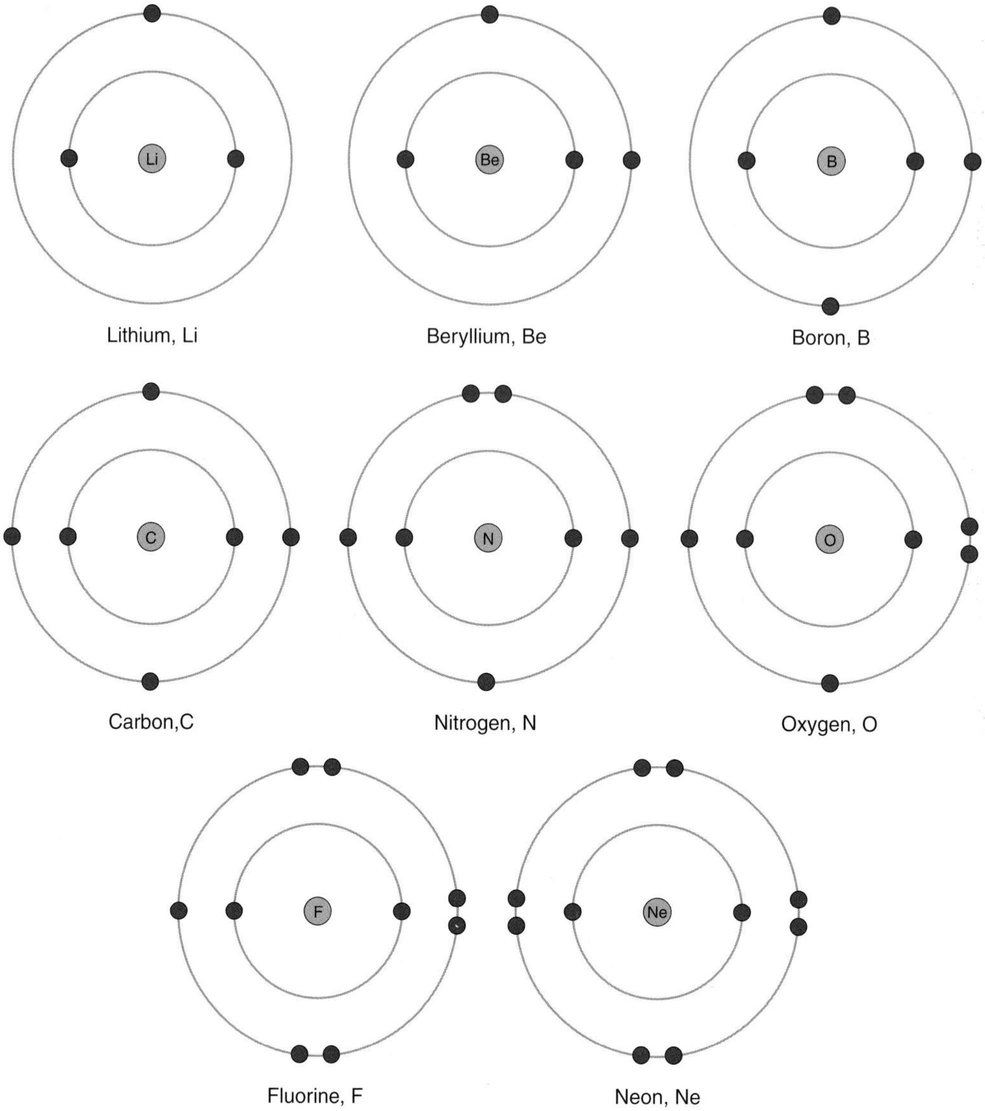

Li — Lithium, Li

Be — Beryllium, Be

B — Boron, B

C — Carbon,C

N — Nitrogen, N

O — Oxygen, O

F — Fluorine, F

Ne — Neon, Ne

The electronic structure of an element can also be shown as a 'dot and cross' diagram. Figure 1.10 shows the electronic structure of fluorine using the 'dot and cross' method.

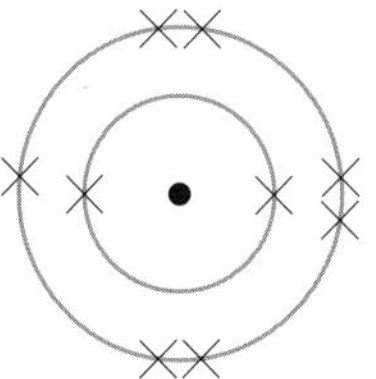

Figure 1.10
The electronic structure of fluorine using the 'dot and cross' method

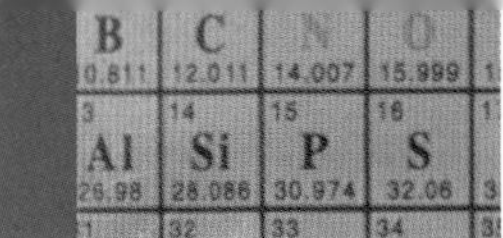

The electronic structures of these elements can be written in a shorter way to show the number of electrons present in each shell. For example, element number 17, chlorine, can be written as $_{17}Cl = 2,8,7$ showing that the 17 electrons in a chlorine atom are arranged as two in the first shell, eight in the second and seven in the third.

Figure 1.11 shows how the electrons are arranged in the first 20 elements.

Figure 1.11

Element	Symbol	Atomic no.	First shell	Second shell	Third shell	Fourth shell
hydrogen	H	1	1			
helium	He	2	2	First shell is now full		
lithium	Li	3	2	1		
beryllium	Be	4	2	2		
boron	B	5	2	3		
carbon	C	6	2	4		
nitrogen	N	7	2	5		
oxygen	O	8	2	6		
fluorine	F	9	2	7		
neon	Ne	10	2	8	Second shell full	
sodium	Na	11	2	8	1	
magnesium	Mg	12	2	8	2	
aluminium	Al	13	2	8	3	
silicon	Si	14	2	8	4	
phosphorus	P	15	2	8	5	
sulphur	S	16	2	8	6	
chlorine	Cl	17	2	8	7	
argon	Ar	18	2	8	8	Third shell full
potassium	K	19	2	8	8	1
calcium	Ca	20	2	8	8	2

Summary

- John Dalton reintroduced after 2000 years the Greek idea that matter was made of solid indivisible particles called atoms.
- The modern model of the atom has a small central **nucleus** made up of **protons** and **neutrons** around which are the **electrons.**
- Protons have a positive charge.
- Neutrons have no charge.
- Electrons have a negative charge and are found in energy layers or shells around the atomic nucleus.
- Atoms have no overall charge because the number of protons equals the number of electrons.
- The **atomic number** (proton number), Z, is the number of protons in the nucleus. It is the same as the number of electrons in shells around the nucleus.
- The **mass number** (nucleon number), A, is the sum of the number of protons and neutrons (nucleons) in the nucleus.
- **Isotopes** are atoms of the same element that have the same number of protons but different numbers of neutrons in their nuclei.
- The electronic structure of an atom of a particular element can be represented using either the 2, 8, 8, 1 or the ‘dot and cross’ convention.
- **Compounds** are substances in which two or more elements are chemically combined.

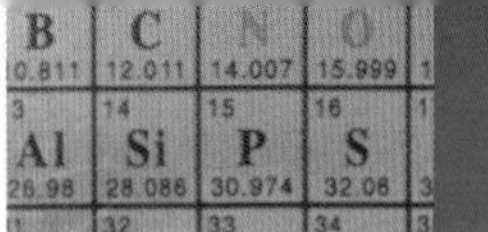

Topic questions

1 a) In what ways did the beliefs of Democritus and Aristotle differ about what matter was made of?
b) Why were Democritus's ideas ignored?
c) John Dalton believed that:
- all elements were made up of very small solid spheres called atoms
- the atoms in a given element are identical and have the same weight
- the atoms of different elements have different weights
- chemical compounds are formed when atoms combine
- a given compound is always made up of the same numbers and types of atoms.

For each of these points explain whether or not they hold true today.

2 Name the three particles present in atoms.

3 Use words from the following list to complete the sentences:

nucleus	**protons**	**shells**	**electrons**
positive	**neutral**	**negative**	**neutrons**

An atom consists of a small central part called the ________ surrounded by several ________. The central part contains ________ which have a ________ charge and ________ which have no charge. The ________ which are found in the shells around the nucleus have a ________ charge.

4 The symbol for an isotope of carbon is $^{14}_{6}C$. How many protons, neutrons and electrons does this form of carbon contain?

5 How are the electrons arranged in the following elements:
a) sulphur S
b) magnesium Mg
c) neon Ne?
d) Which one of these elements is a noble gas?

6 Draw a diagram to show how all the particles are arranged in an atom of $^{31}_{15}P$.

7 What is the maximum number of electrons that can be held in
a) shell 1?
b) shell 2?
c) shell 3?

1.2 Bonding

Co-ordinated	Modular
11.2	08 (10.2)

Metals and non-metals

When elements are sorted into metals and non-metals it can be seen that their electronic structures have some noticeable similarities and differences.

Figure 1.12

The electronic structures of selected elements			
Metals	**Electron structure**	**Non-metals**	**Electron structure**
lithium	2,1	carbon	2,4
beryllium	2,2	nitrogen	2,5
sodium	2,8,1	oxygen	2,6
magnesium	2,8,2	fluorine	2,7
aluminium	2,8,3	phosphorus	2,8,5
potassium	2,8,8,1	sulphur	2,8,6
calcium	2,8,8,2	chlorine	2,8,7

The first thing to notice is that the metals have either one, two or three electrons in their outer shell, but non-metals have four, five, six or seven electrons in theirs. All elements react so that their electronic structures become identical to that of the noble gas nearest to them in the periodic table (i.e. they try to get a full outer shell of electrons). In this way the elements become more stable (see section 5.3).

Since metals and non-metals have quite different structures, they behave differently as they try to achieve the electronic structure of the nearest noble gas. Metals usually react by losing their outer electrons but non-metals try to gain more outer electrons to complete their outer shell. There are three main ways that elements can do this. Each method results in a chemical bond being formed. The methods produce **covalent bonds**, **ionic bonds** or **metallic bonds**.

Molecules

The only elements that exist in nature as individual atoms are the noble gases. This is because their outermost electron shells are already full and stable. The atoms of all other elements join together in some way to form molecules and the method used depends on the type of element they are and the type of element with which they join.

Remember: Atoms join together to make their electronic structures more stable. Each element tries to achieve the same electronic structure as the noble gas nearest to it in the periodic table.

Covalent bonds in elements

Covalent bonds are formed when atoms *share* electrons so that they all end up with a stable, noble gas structure in which all the electron shells are full. If the elements are non-metal elements, their outer shells are short of electrons and they achieve full shells by covalent bonding.

Hydrogen molecules, H_2

Hydrogen atoms have only one electron in their outer shell. The nearest noble gas to hydrogen is helium which has two outer electrons. Two hydrogen atoms join by sharing their electrons so that each has control over two (see Figure 1.13).

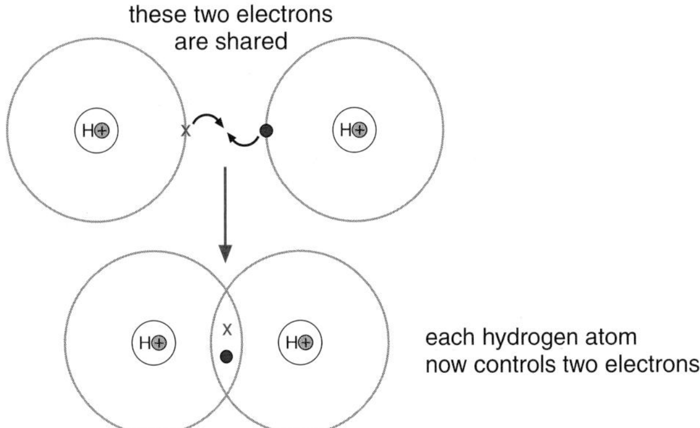

Figure 1.13
How two hydrogen atoms combine to make a hydrogen molecule

This produces a **molecule** of hydrogen containing two hydrogen atoms chemically joined together. It is a diatomic (two atom) molecule and the two atoms are joined by a pair of shared electrons. This is a covalent bond.

Sometimes the sharing of one pair of electrons is not enough to give all the atoms present completely full shells but elements always adapt so that stable structures are produced. Figure 1.14 shows how two oxygen atoms join together to make an oxygen molecule (O_2) by forming a double covalent bond. Each oxygen atom has gained control over eight electrons in its outer shell.

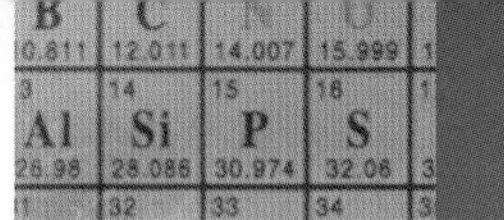

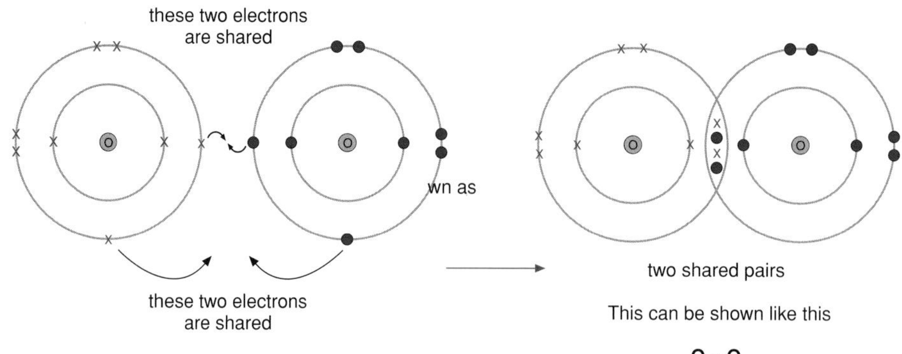

Figure 1.14
How two oxygen atoms combine to form an oxygen molecule with a double covalent bond

Covalent bonds in compounds

Covalent bonds can also be formed between atoms of different elements. In each case the atoms are linked by pairs of shared electrons – each atom usually donates one electron to each shared pair.

Figures 1.15 to 1.17 show how covalent bonds are formed in molecules of methane (CH_4), ammonia (NH_3) and hydrogen chloride (HCl), respectively.

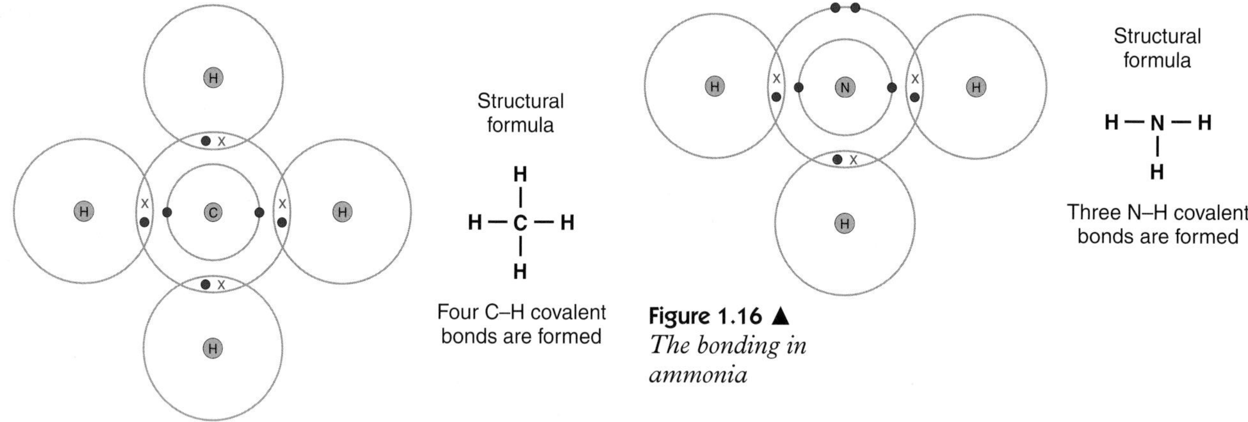

Figure 1.15▼
The bonding in methane

Figure 1.16 ▲
The bonding in ammonia

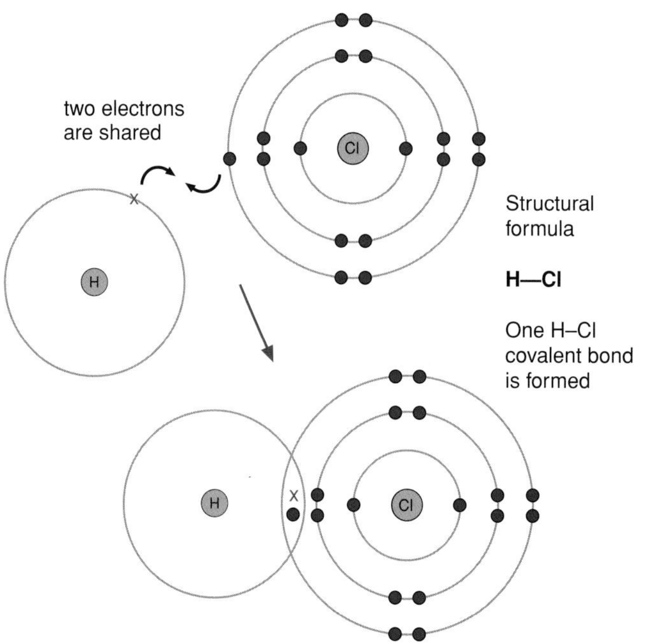

Figure 1.17
The bonding in hydrogen chloride

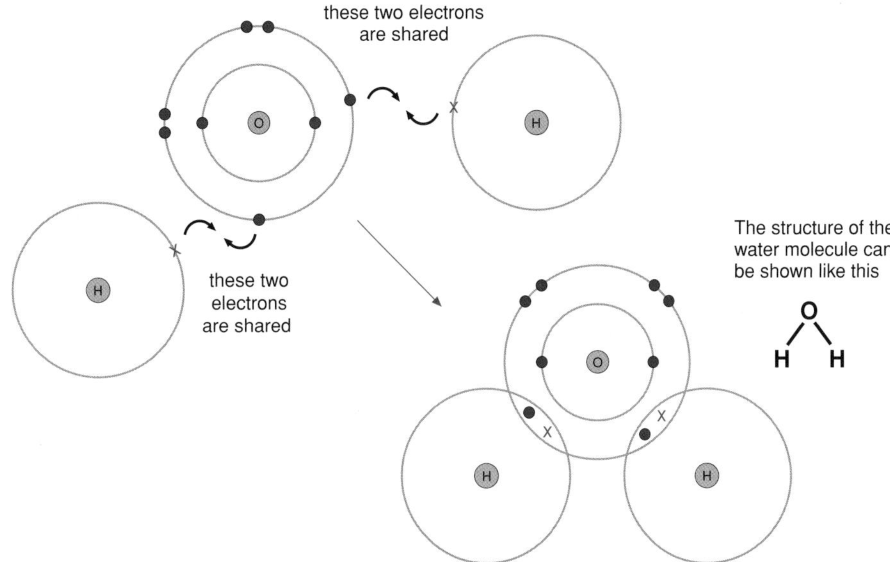

Figure 1.18
How two hydrogen atoms and one oxygen atom form a water molecule

Here are some melting and boiling points for some familiar covalently-bonded materials.

Figure 1.19

Material	Formula	Melting point (°C)	Boiling point (°C)	Normal state at room temperature
carbon dioxide	CO_2	sublimes* at −78.5		gas
carbon monoxide	CO	−199	−192	gas
methane	CH_4	−182	−162	gas
ethane	C_2H_6	−183	−88	gas
water	H_2O	0	100	liquid

*The solid turns directly to a gas without melting to a liquid first

Notice how covalent compounds only contain non-metal elements and that they are usually gases, liquids or low melting point solids. The covalent bond in each molecule is a strong bond.

The reason covalently-bonded compounds have such low melting and boiling points is because the forces *between* neighbouring covalent molecules are very weak.

Covalent compounds do not usually dissolve in water nor do they conduct electricity because they do not contain charged particles.

Important points to remember about covalent bonds

- They are formed only between non-metal elements.
- They are formed by sharing pairs of electrons – each atom donating one electron to each shared pair.
- They produce clusters of atoms chemically joined together, known as molecules.
- Whilst the atoms in the molecules are strongly joined together, the forces between molecules are quite weak so covalent compounds are often gases (e.g. oxygen and hydrogen), low boiling point liquids (e.g. water and petrol) or low melting point solids (candle wax).
- They are frequently insoluble in solvents like water.
- They do not conduct electricity in either the solid or molten state.

Giant molecules

Some covalent materials behave in a different way. For example diamond and graphite have high melting points (more than 3500°C) and sand (silicon dioxide) melts at 1610°C. This is because they are not simple little molecules but have giant three-dimensional structures (see Figure 1.20).

Figure 1.20 *The arrangement of covalently bonded carbon atoms in (a) diamond and (b) graphite*

(a)

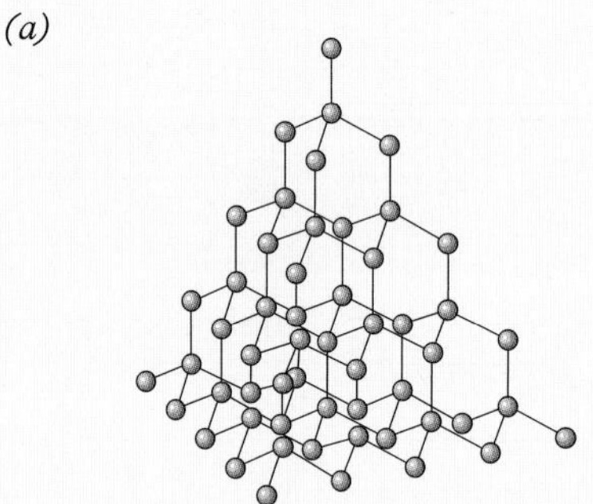

In diamond each carbon atom forms four covalent bonds which produce a rigid, giant covalent structure.

(b)

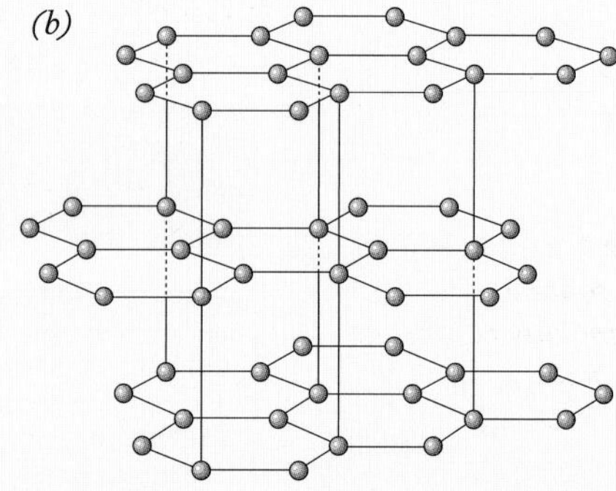

In graphite each carbon atom forms three covalent bonds. The carbon atoms form layers which are free to slide over each other. In graphite there are free electrons which allow the graphite to conduct electricity.

Did you know?

Diamond and graphite are pure forms of the same element, carbon. Diamond is the hardest naturally-occurring substance yet graphite is soft enough to leave a grey mark on paper and is used in pencil lead.

Ionic bonds

Ionic bonds are formed when electrons are transferred (moved) from one atom to another. The electronic structures of non-metals show that they are generally short of electrons, but the opposite is true for metals. Metals have one, two or three electrons outside a full electron shell and they react by losing these electrons to other elements. When metals combine with non-metals the electrons in the outer shell of the metal atoms transfer completely to the outer shell of the non-metal atoms.

This means that each 'atom' becomes charged. These charged 'atoms' are called **ions**. Positively-charged ions are called **cations** and negatively-charged ions are called **anions**. In atoms there are equal numbers of positively-charged protons and negatively-charged electrons so the overall charge is zero.

In ions this is not the case. A sodium atom (Na = 2,8,1) contains 11 protons and 11 electrons. If it loses its outer electron, it still has 11 protons but now has only 10 electrons so it becomes a positively-charged ion.

$$\text{Na (2,8,1)} - 1e^- \rightarrow \text{Na}^+ \text{ (2,8)}$$

Note that the sodium ion (Na^+) has the same electronic structure as the noble gas neon, Ne (2,8).

A chlorine atom contains 17 protons and 17 electrons but if it gains an extra electron into its outer shell to make it full, it will have 17 protons and 18 electrons and becomes a negatively-charged ion.

$$\text{Cl (2,8,7)} + 1e^- \rightarrow \text{Cl}^- \text{ (2,8,8)}$$

Note that the chlorine ion (Cl^-) has the same electronic structure as the noble gas argon, Ar (2,8,8).

When sodium and chlorine join together, electrons transfer from the outer shells of sodium atoms to the outer shells of chlorine atoms and lots of Na^+ and Cl^- ions are formed. Since these ions are of opposite charge they attract strongly and a strong ionic bond is formed (see Figure 1.21).

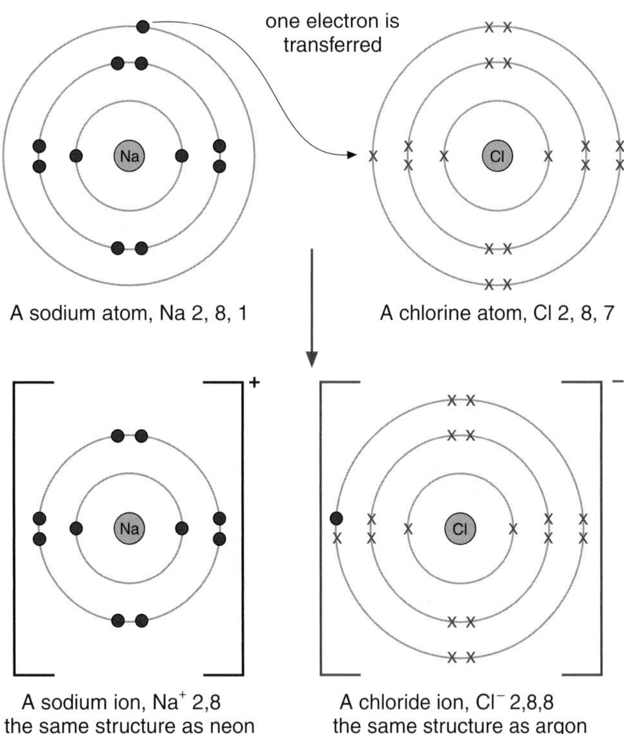

Figure 1.21
How sodium and chlorine atoms form an ionic bond

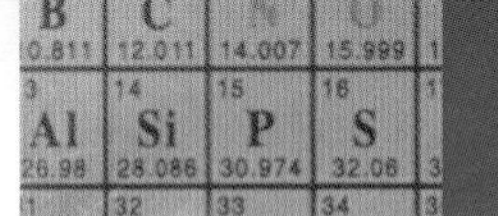

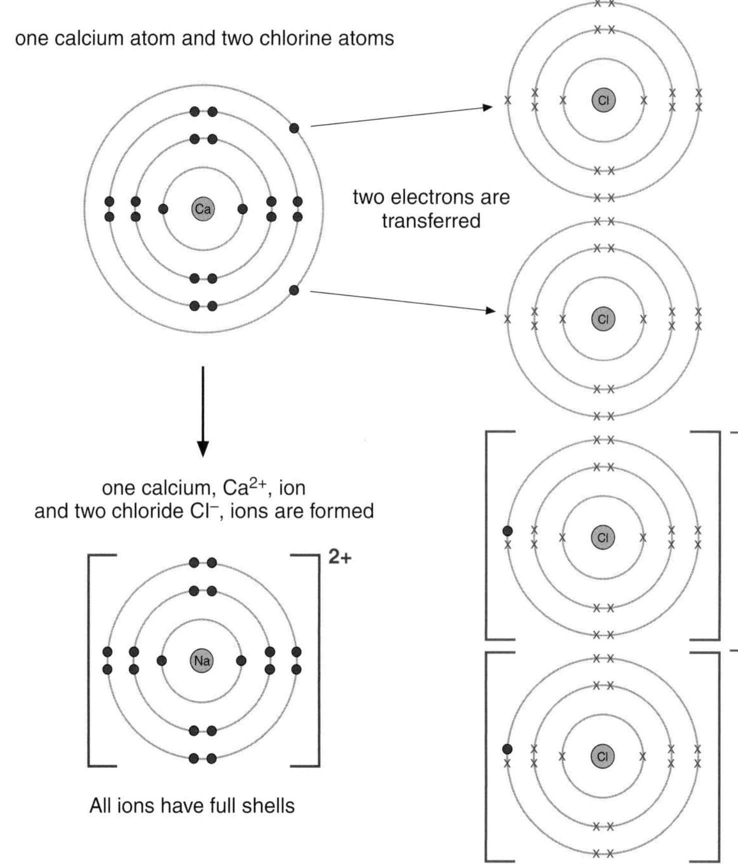

Figure 1.22
How the ionic bond is formed in calcium chloride

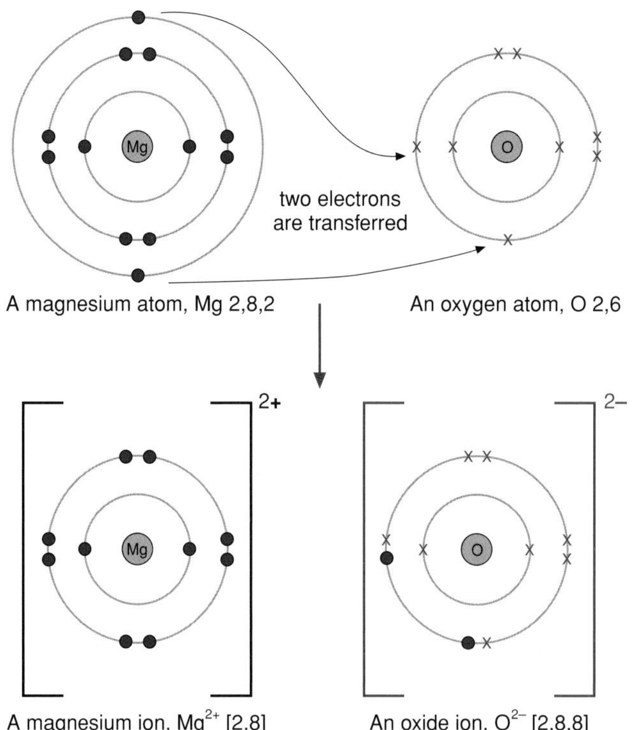

Figure 1.23
How the ionic bond is formed in magnesium oxide

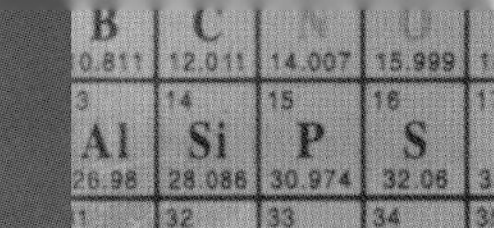

The electronic structures of other common ions are shown in Figure 1.24.

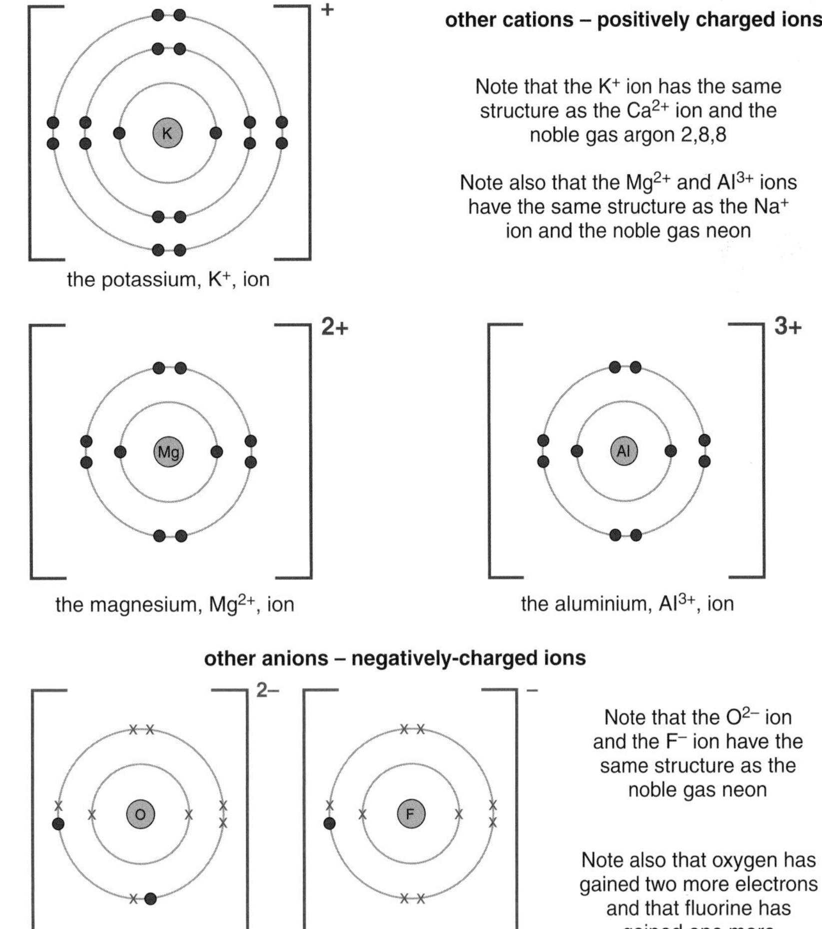

Figure 1.24 *The electronic structures of some other ions*

Here are some melting points and boiling points of familiar ionic compounds. Note that every one contains both a metal and a non-metal element.

Figure 1.25

Material	Formula	Melting point (°C)	Boiling point (°C)	Normal state at room temperature
sodium chloride	$NaCl$	801	1465	solid
copper oxide	CuO	1326	very high	solid
iron oxide	Fe_2O_3	1565	very high	solid
magnesium oxide	MgO	2800	3600	solid
potassium chloride	KCl	770	1407	solid
lead bromide	$PbBr_2$	373	916	solid
aluminium oxide	Al_2O_3	2045	2980	solid

All ionic compounds are solids with high melting points. This is completely different from covalent compounds that have low melting and boiling points.

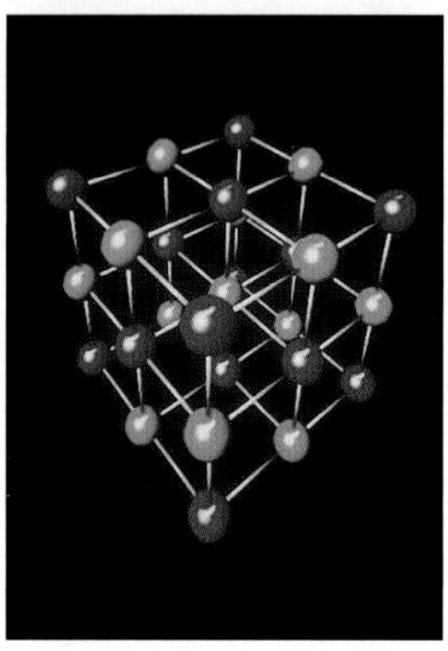

Figure 1.26
The arrangement of sodium (Na^+) ions and chloride (Cl^-) ions in sodium chloride

More about ionic bonding

In ionic compounds the charges on the ions will attract any ion that has the opposite charge. Each positive ion becomes surrounded by and attracted to lots of negative ions and each negative ion becomes surrounded by and attracted to lots of positive ions. This can be seen in Figure 1.26 which shows the arrangement of the Na^+ and Cl^- ions in sodium chloride (NaCl). The result is a large three-dimensional structure.

Melting this solid would involve breaking the ionic bonds between all the oppositely-charged ions to release them to become mobile. A lot of energy is required to do this so the melting points of ionic solids are high.

Ionic compounds

Ionic compounds are solids with giant structures but many of them will dissolve in water.

Ionic solids will not conduct electricity when in the solid state. They will, however, conduct electricity when they are molten or dissolved in water. In the solid the ions cannot move (see Figure 1.26) but in the molten state or in solution they can. Mobile ions can carry the current through the liquid. The positively-charged ions move towards the negative electrode and the negatively-charged ions move towards the positive electrode. This two-way flow of ions towards electrodes of opposite charge is known as **electrolysis**. Electrolysis results in some form of chemical change (see Figure 1.27).

Figure 1.27
How the ions move in a liquid ionic compound

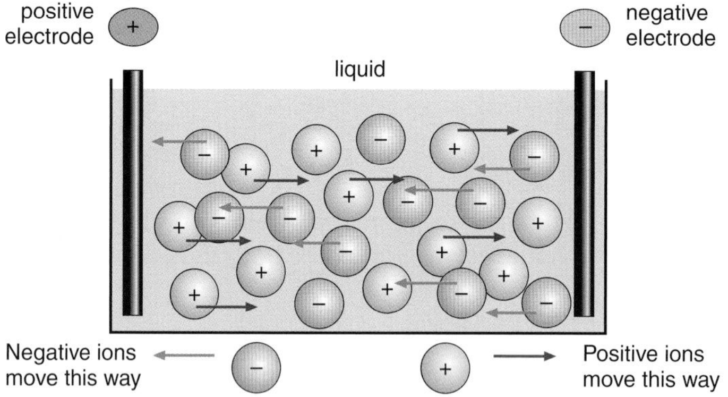

Important points to remember about ionic bonds

- They are formed only between metals and non-metal elements.
- They are formed by the transfer of electrons from metal atoms to non-metal atoms.
- They produce charged particles, known as ions.
- The ionic bond is a very strong bond formed as a result of very strong electrostatic forces of attraction between lots of oppositely-charged ions.
- Ionic compounds are always solids with high melting points.
- Ionic solids are often soluble in water.
- Ionic solids do not conduct electricity but when molten or in aqueous solution they do.

Metals and the metallic bond

With the exception of mercury, all metals are solids at room temperature and most of them have high melting points. Metals are not soluble in water, though some of them react with water and appear to dissolve. Metals are good conductors of electricity.

Metal atoms have either one, two or three electrons in their outer shell. To get a stable noble gas structure the metal atoms need to lose these electrons. When metal atoms are together in a piece of metal, these outer electrons become 'pooled' together to form a 'sea' of **free (mobile) electrons**. The nuclei of the metal atoms and all their completely full electron shells occupy a fixed position within a giant three-dimensional crystal structure and the sea of free, mobile electrons is able to flow through this structure giving metals their characteristic properties. Imagine that a box is filled with marbles to represent the nuclei and full shells of the atoms and then water (representing the sea of outer electrons) is poured into the box to cover the marbles. The marbles will be arranged in a neat ordered pattern but the water will be free to flow between them (see Figure 1.28).

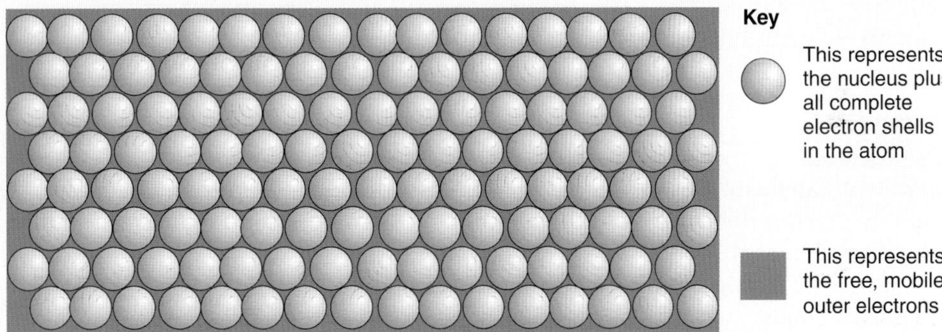

Figure 1.28 *How the atoms and outer electrons are arranged in a metal*

The free, mobile electrons are responsible for metals being conductors of heat and electricity. The metal atoms are tightly packed together bonded by the sea of free electrons. This means that individual atoms are not easy to separate so metals are strong, hard and have high melting points.

Metals are both malleable and ductile which means that they can be bent into a different shape or can be drawn out into wires. This is because the free electrons allow the tightly packed atoms to move into different positions within the crystal structure (see Figure 1.29).

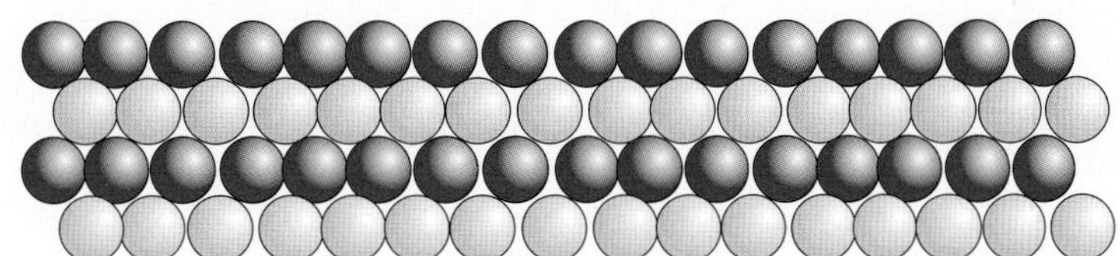

The atom arrangement in a straight piece of metal

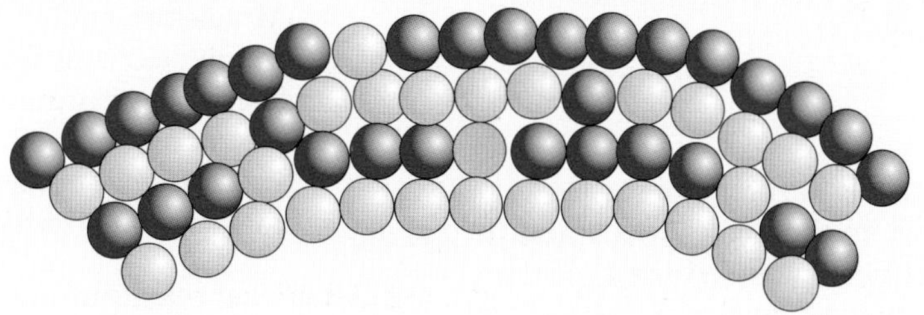

The atom arrangement when the metal is bent – some atoms are forced into different layers

Figure 1.29 *What happens when a metal bends*

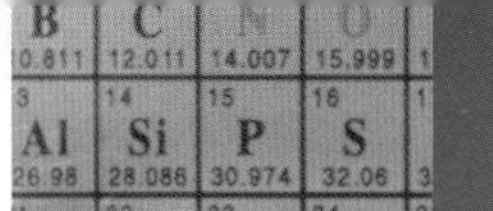

Getting together

The flow chart in Figure 1.30 can be used to work out the type of chemical bond likely to be present in a substance.

Figure 1.30

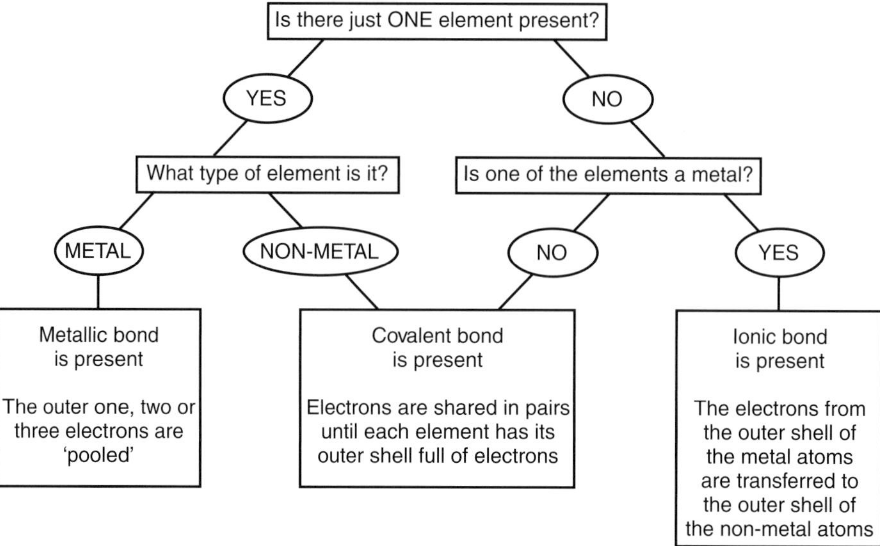

Summary

- When non-metal elements join together they do so by sharing electrons from the highest occupied energy level (outermost occupied shell). The strong bonds formed are called **covalent bonds.**
- Covalent compounds consist of molecules
- Covalent compounds have low melting points, low boiling points and do not conduct electricity.
  - Although the covalent bonds that hold a molecule of a covalent compound together are strong, the intermolecular forces are weak.
  - The molecules of a covalent compound carry no overall electric charge.
- Covalent compounds can be represented using either the 'dot and cross' or structural formula conventions.
  - Giant covalent structures have high melting points because of the large number of strong covalent bonds in their structure (lattice). Diamond and graphite and silicon dioxide are examples of giant covalent structures.
- Metals and non-metals join together when electrons transfer from the highest occupied energy level (outermost occupied shell) of the metal atom to the highest occupied energy level (outermost occupied shell) of the non-metal atom forming an **ionic bond.**
- The gaining or losing of electrons produces ions which have the electronic structure of a noble gas.
- Ionic compounds are all solids with high melting points, high boiling points and they conduct electricity when molten or in solution.
- Ions and ionic compounds can be represented using either the 'dot and cross' or the 2,8,8,1 conventions.
  - Ionic compounds form giant ionic lattices in which there are strong forces of attraction between the oppositely charged ions.
  - Ionic compounds have high melting points and high boiling points because of the strong forces of attraction.
  - When ionic compounds are melted or dissolved in water the ions become free to move and so the ionic compounds conduct electricity.
  - Metals are made of giant structures in which electrons from the highest occupied energy level (outermost shell) of the metal atom are free to move through the whole structure. These **free electrons** allow the structure to conduct electricity and heat.
- Chemical reactions involve the rearrangement of atoms to make new substances.

Topic questions

1 Use the copy of the periodic table (on page 82) to identify each of the following elements as either metal or non-metal and then decide the type of bond that would be formed between each pair.
a) magnesium and fluorine
b) oxygen and sulphur
c) zinc and chlorine
d) phosphorus and oxygen

2 Show how the elements sulphur, ${}_{16}S$, and hydrogen, ${}_{1}H$, can combine together to make the compound hydrogen sulphide, H_2S.

3 Show how the elements lithium, ${}_{3}Li$, and fluorine, ${}_{9}F$, can combine together to make the compound lithium fluoride, LiF.

4 Show how sodium, ${}_{11}Na$, and oxygen, ${}_{8}O$, can combine to form the compound sodium oxide, Na_2O.

5 The properties of substances often indicate the type of bond present in them. Study the following descriptions of substances A, B, C and D and say what type of chemical bond is likely to be present.

Description of compound	Type of bond present
A is a white solid that melts at 960°C. It dissolves in water and conducts an electric current when in solution and when molten.	
B is a liquid which does not dissolve in water. Its boiling point is 87°C and it burns easily.	
C is a blue crystalline solid that dissolves easily in water. The resulting solution conducts electricity. The solid decomposes before it melts.	
D is a gas with a very strong smell. It is very soluble in water and the resulting solution conducts electricity. If the gas is cooled until it changes into a liquid, this liquid does not conduct electricity.	

6 Use the flow chart in Figure 1.30 to work out the type of chemical bond present in each of the following substances.
a) magnesium, b) fluorine, c) sodium sulphide, d) lithium oxide, e) nitrogen, f) hydrogen sulphide, g) phosphine (a compound of phosphorus and hydrogen)

1.3	
Co-ordinated	Modular
11.8	07 (10.10)

Quantitative chemistry

Quantitative chemistry is concerned with the amounts of materials involved in chemical reactions. In some chemical reactions it looks as if the mass has changed. For example when solutions of lead nitrate and potassium iodide are mixed, a dense yellow solid (called a precipitate) of lead iodide is formed. Although it appears that the mass has increased, it hasn't (see Figure 1.31).

Getting together

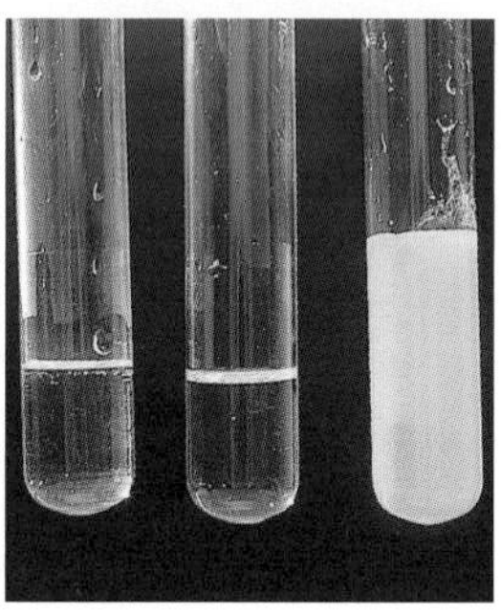

Figure 1.31
The formation of solid lead iodide (right) by mixing solutions of lead nitrate and potassium iodide (left)

It is important to remember that **during a chemical change, matter (material) is neither created nor destroyed**. This is a fundamental law in science known as **the Law of Conservation of Mass**. During a chemical reaction atoms are neither created nor destroyed, they are simply rearranged to form new substances. In some reactions gases are produced, for example when magnesium is added to dilute acid in a flask. Gases have mass so the mass of the apparatus will decrease if the gas is allowed to escape into the air.

Chemical calculations are based on the Law of Conservation of Mass. In chemical reactions it is the number of atoms and molecules taking part that is important. In a bank the cashier 'counts' coins by 'weighing' them. The same can be done in chemistry. It is necessary to take into account that different atoms have different masses. This is done using the quantity known as the **relative atomic mass (A_r)**.

Did you know?

The element chlorine $_{17}Cl$ has two isotopes – ^{35}Cl and ^{37}Cl. Both forms of chlorine have 17 protons in their nucleus and 17 electrons in shells around the nucleus but ^{35}Cl has 18 neutrons in its nucleus whilst ^{37}Cl has 20. The ratio of ^{35}Cl atoms to ^{37}Cl atoms is 3 : 1. This gives an average value of 35.5 for the relative atomic mass of chlorine.

Relative atomic mass does not have any units because it is a ratio and not a real mass. The relative atomic mass is approximately the number of times one atom of the element is heavier than one atom of hydrogen. One atom of chlorine is (on average) 35.5 times heavier than one atom of hydrogen.

Some commonly used relative atomic masses are shown in Figure 1.32.

Figure 1.32

Element	Relative atomic mass	Element	Relative atomic mass
hydrogen, H	1	sodium, Na	23
chlorine, Cl	35.5	magnesium, Mg	24
carbon, C	12	calcium, Ca	40
oxygen, O	16	aluminium, Al	27
nitrogen, N	14	lead, Pb	207
sulphur, S	32	iron, Fe	56

Did you know?

Relative atomic masses used to be calculated by comparing the mass of an atom to the mass of a hydrogen atom and calling the relative atomic mass of hydrogen 1. For practical reasons carbon is now used as the basis for calculating relative atomic masses. The mass of an atom of an element is compared to the mass of the ^{12}C isotope of carbon on a scale on which the ^{12}C isotope has a mass of 12 atomic mass units.

Calculating relative molecular masses and relative formula masses

Most substances exist either as molecules or ions so it is important to be able to compare the masses of these particles with those of other substances. The **relative formula mass (M_r)** of a simple one element ion (e.g. the Cl^- ion) is the same as the relative atomic mass of the atom from which it was formed because the mass of an electron is so small. The relative mass of the Cl^- ion is therefore 35.5.

Where substances consist of molecules, the relative mass of the molecule is obtained by adding together the relative masses of all the atoms within the molecule. For example, water is H_2O, indicating that it consists of two atoms of hydrogen (H = 1) and one atom of oxygen (O = 16).

The relative molecular (or formula) mass of water is $(2 \times 1) + 16 = 18$.

The relative molecular mass of sulphuric acid (H_2SO_4) is $[(2 \times 1) + (32) + (4 \times 16)] = 98$.

Calculating the percentage of an element in a compound

To do this you must first calculate the relative formula mass of the compound and then calculate the percentage of the element.

Example: What is the percentage of carbon (C) in sodium carbonate (Na_2CO_3)?

Na = 23, C = 12, O = 16

The relative formula mass of Na_2CO_3 is 46 + 12 + 48 = 106.

Carbon makes up just 12 parts of this so:

the percentage of carbon in Na_2CO_3 is $\frac{12}{106} \times 100 =$ **11.43%**

The mole

It is rarely necessary to know the exact number of particles being weighed out. What is needed is the relative number being used. This is important when making new compounds from existing ones so that when the substances are mixed they are all used up and there is no unreacted material left.

Just as shops sell eggs in fractions or multiples of a dozen, so the chemist measures out chemicals in fractions or multiples of a **mole**. The mole is the number of atoms in 1 g of hydrogen (H = 1). It is also the number of atoms in 12 g of carbon (C = 12) or 23 g of sodium (Na = 23) etc.

If the relative mass of a substance is known, then 1 mole of that substance is its relative mass in grams. For example, the relative mass of water is 18 (see earlier) so 1 mole of water is 18 g.

The relative mass of sulphuric acid is 98 so 1 mole of it has a mass of 98 g.

Did you know?

The number of atoms or molecules in a mole is 6×10^{23}. That is 600 000 000 000 000 000 000 000 or six hundred thousand million, million, million!

If hydrogen atoms were laid side by side in a row, there would be 300 million of them in 1 cm!

If one mole of hydrogen atoms were laid side by side, the line would be 20 000 000 000 km long. It would take a ray of light about 18 hours to travel that distance!

Calculating reacting masses

We can use the previous information to calculate the amounts of materials used in chemical reactions.

Example 1: How much copper oxide (CuO) can be obtained by heating 12.4 g of copper carbonate ($CuCO_3$) until there is no further change?

Cu = 64, C = 12, O = 16

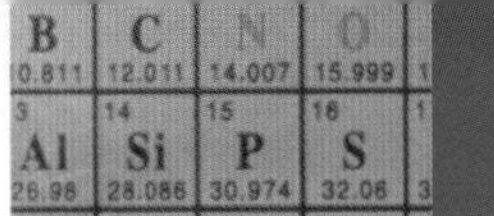

First write the equation for the reaction:

$$CuCO_3(s) \rightarrow CuO(s) + CO_2(g)$$

This tells us that 1 mole of $CuCO_3$ produces 1 mole of CuO and 1 mole of CO_2.

So, 1 mole $CuCO_3$ produces 1 mole CuO
(64 + 12 + 48 = 124 g) $CuCO_3$ produce (64 + 16 = 80 g) CuO
124 g $CuCO_3$ produce 80 g CuO
12.4 g $CuCO_3$ produce **8.0 g** CuO

Example 2: What mass of sodium carbonate (Na_2CO_3) is needed to react with and neutralise a solution containing 5 g of nitric acid (HNO_3)?

Na = 23, C = 12, O = 16, H = 1, N = 14

Starting with the 'balanced' equation:

$$2HNO_3(aq) + Na_2CO_3(s) \rightarrow 2NaNO_3(aq) + H_2O(l) + CO_2(g)$$

2 moles of HNO_3 react with 1 mole of Na_2CO_3

[1 mole of HNO_3 is 1 + 14 + 48 = 63 g so 2 moles is 126 g]

[1 mole of Na_2CO_3 is 46 + 12 + 48 = 106 g]

So, 126 g HNO_3 react with 106 g Na_2CO_3

1 g HNO_3 reacts with $\frac{106}{126}$ g of Na_2CO_3

5 g HNO_3 react with $5 \times \frac{106}{126}$ g of Na_2CO_3 = **4.21 g** Na_2CO_3

Reactions involving gases

Quantities of gas are measured by volume instead of mass. Experiments show that one mole of any gas measured at the same temperature and pressure has the same volume. Since gases are easy to compress, their volumes must always be compared under the same conditions of temperature and pressure.

Under standard temperature and pressure conditions (0°C and 1 atmosphere pressure) the volume of 1 mole of gas is 22 400 cm^3 (22.4 dm^3). This is called the **molar volume**. Under normal room temperature and pressure conditions it is about 24 dm^3.

This value is used in calculations.

Example 3: Calculate the volume of hydrogen gas produced under room conditions when 8 g of calcium reacts with water.

Ca = 40, H = 1, O = 16

1 mole of gas occupies 24 dm^3 under normal room conditions

The equation for the reaction is: $Ca(s) + 2H_2O(l) \rightarrow Ca(OH)_2(aq) + H_2(g)$

1 mole Ca produces 1 mole H_2

40 g Ca produce 24 dm^3 H_2

8 g Ca produce $\left(\frac{24}{40}\right) \times 8$ dm^3 H_2 = **4.8 dm^3** H_2

Moles and solutions

When solutions are used, their concentrations are shown in one of two ways: a) g/dm^3 or b) mol/dm^3. (1 dm^3 is 1000 cm^3. The dm^3 is the SI equivalent to the litre.)

The concentration of a solution in moles per dm^3 is called the molarity of the solution. For example, the relative formula mass of sodium hydroxide (NaOH) is (23 + 16 + 1 = 40) so a solution containing 1 mole of sodium hydroxide per dm^3 would contain 40 g of sodium hydroxide per dm^3 and would be known as a 1 molar solution (1 M solution). A solution containing 20 g of sodium hydroxide per dm^3 would be a 0.5 M solution. A solution containing 80 g sodium hydroxide per dm^3 (or 40 g in 500 cm^3) would be a 2.0 M solution.

Remember that solutions of the same molarity contain the same number of 'molecules' of the dissolved substance.

Changing grams to moles and moles to grams

To change grams to moles use:

$$\text{amount of substance (mol)} = \frac{\text{mass of substance (g)}}{\text{molar mass (g/mol)}}$$

Example 4: How many moles are there in 12.6 g of nitric acid (HNO_3)?

H = 1, N = 14, O = 16

1 mole of HNO_3 is 1 + 14 + 48 = 63 g

Using the formula: amount of substance (mol) $= \frac{12.6}{63} = 0.2$ mole

To change moles to grams use:

mass of substance (g) = amount of substance (mol) × molar mass (g/mol)

Example 5: What is the mass of 4 moles of copper sulphate ($CuSO_4$)?

Cu = 64, S = 32, O = 16

1 mole of $CuSO_4$ is 64 + 32 + 64 = 160 g

Using the formula: mass of substance (g) = 4 × 160 = **640 g** $CuSO_4$

Calculating the formula of a compound

When iron (Fe) reacts with chlorine (Cl_2) 1.12 g of iron produce 3.25 g of iron chloride. The formula of iron chloride can be deduced by first calculating the amount of each element present in the compound (Fe = 56, Cl = 35.5).

If 3.25 g of iron chloride are formed from 1.12 g of iron the mass of chlorine present must be 3.25 − 1.12 = 2.13 g

1.12 g of iron $= \frac{1.12}{56} = 0.02$ mole of iron

2.13 g of chlorine $= \frac{2.13}{35.3} = 0.06$ mole of chlorine

So 0.02 mole of iron combines with 0.06 mole of chlorine

Therefore 1 mole of iron combines with $\frac{0.06}{0.02} = 3$ mole of chlorine

The formula for the iron chloride is therefore $FeCl_3$.

Summary

- During a chemical reaction matter is neither created nor destroyed.
- The **relative atomic mass** of an element is the mass of one atom of the element compared to the mass of one atom of hydrogen.
- The **relative molecular mass/relative formula mass** of a substance is found by adding together the relative atomic masses of all the atoms in one molecule of it.
- It is possible to calculate the relative formula mass of a compound when the formula is provided.
- It is possible to calculate the percentage of an element in a compound when the formula is provided.
- It is possible to calculate masses/volumes of reactants or products if the balanced symbol equations or data regarding the volumes or masses of some of the reactants or products are provided.
- It is possible to calculate the ratios of elements in a compound and hence the formula if the masses or percentage composition is provided.

Topic questions

Use the following relative atomic masses in any calculations that follow.

H = 1, C = 12, O = 16, N = 14, Cl = 35.5, Na = 23, Mg = 24, S = 32, Pb = 207, Ag = 108, K = 39

1 mole of gas occupies 24 dm^3 under room conditions

1 Calculate the relative formula masses of each of the following substances:

SO_2 $MgSO_4$ $(NH_4)_2SO_4$ $NaNO_3$

2 Which of the following ammonium compounds contains the greater percentage of nitrogen (N)?

Ammonium nitrate, NH_4NO_3, or ammonium carbonate, $(NH_4)_2CO_3$

3 Calculate the mass of each of the following:
a) 1 mole of nitrogen dioxide (NO_2)
b) 3 moles of magnesium carbonate ($MgCO_3$)
c) 0.2 mole of lead nitrate ($Pb(NO_3)_2$)

4 How many moles are present in each of the following?
a) 24 g of magnesium sulphate ($MgSO_4$)
b) 2.8 g of potassium hydroxide (KOH)
c) 48 dm^3 of carbon dioxide (under room conditions)

5 When sodium chloride solution (NaCl(aq)) is added to silver nitrate solution ($AgNO_3$(aq)) a white precipitate of silver chloride (AgCl(s)) is formed according to the following equation:

$$NaCl(aq) + AgNO_3(aq) \rightarrow AgCl(s) + NaNO_3(aq)$$

What mass of silver chloride could be obtained from a solution that contained 3.4 g of silver nitrate?

6 What volume of carbon dioxide gas (under room conditions) could be obtained from the action of an excess of dilute sulphuric acid (H_2SO_4(aq)) on 16.8 g of sodium hydrogencarbonate ($NaHCO_3$)?

$$2NaHCO_3(s) + H_2SO_4(aq) \rightarrow Na_2SO_4(aq) + 2H_2O(l) + 2CO_2(g)$$

Examination questions

1 The diagrams below represent three atoms, **A**, **B** and **C**.

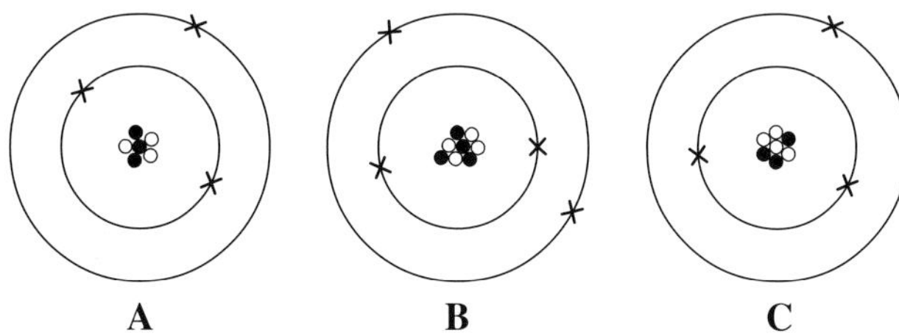

a) Two of the atoms are from the **same** element.
   i) Which of **A**, **B** and **C** is an atom of a different element? *(1 mark)*
   ii) Give one reason for your answer. *(1 mark)*

b) Two of these atoms are isotopes of the same element.
   i) Which **two** are isotopes of the same element?
   ii) Explain your answer. *(3 marks)*

2 This question is about elements and atoms.

a) About how many different elements are there?

40 60 80 100 200 *(1 mark)*

b) The following are parts of an atom:

electron neutron nucleus proton

Choose from the list the one which:
   i) has no electrical charge;
   ii) contains two of the other particles;
   iii) has very little (negligible) mass. *(3 marks)*

c) Scientists have been able to make new elements in nuclear reactors. One of these new elements is fermium. An atom of fermium is represented by the symbol below.

$$^{257}_{100}\mathrm{Fm}$$

   i) How many protons does this atom contain?
   ii) How many neutrons does this atom contain? *(2 marks)*

3 a) Atoms are made of sub-atomic particles. Complete the **six** spaces in the table.

Name of sub-atomic particle	Relative mass	Relative charge
	negligible	
Neutron		
	1	

(3 marks)

b) Complete the spaces in the sentences.
   i) The atomic number of an atom is the number of _______ in its nucleus and is equal to the number of _______ if the atom is not charged. *(1 mark)*
   ii) The mass number of an atom is the total number of _______ and _______ in its nucleus. *(1 mark)*

c) The table gives information about the atoms of three elements.

Name of element	Chemical symbol	Number of electrons in: 1st shell	2nd shell	3rd shell
Fluorine	F	2	7	0
Neon	Ne	2	8	0
Sodium	Na	2	8	1

Two of these elements can react together to form a chemical compound.
   i) What is the name and the formula of this compound? *(2 marks)*
   ii) What type of bonding holds this compound together? *(1 mark)*
   iii) Explain, in terms of electron transfer, how the bonding occurs in this compound. *(2 marks)*

4 a) i) Ammonium nitrate is one type of artificial fertiliser. Calculate the relative formula mass of ammonium nitrate NH_4NO_3. (Relative atomic masses: H = 1, N = 14, O = 16.) *(1 mark)*
   ii) Use your answer to part (a)(i) to help you calculate the percentage by mass of nitrogen present in ammonium nitrate NH_4NO_3. *(2 marks)*

b) One compound of vanadium is vanadium oxide.
A sample of vanadium oxide contained 10.2 g of vanadium and 8.0 g of oxygen.
Calculate the formula of this vanadium oxide.
You must show **all** your working to gain full marks.
(Relative atomic masses: V = 51, O = 16). *(3 marks)*

Chapter 2

Representing reactions

Key terms

chemical equation • compound • displacement reaction • element • endothermic reaction • exothermic reaction • ions • mixture • neutralisation • oxidation • periodic table • precipitate • precipitation • product • reactant • reduction • reversible reaction • thermal decomposition

2.1	
Co-ordinated	Modular
11.7	07/08 (10.9)

Representing chemical symbols, formulae and reactions

Elements

All matter is made up from a limited number of simple substances called **elements**. There are 94 naturally-occurring elements of which 11 are gases, two are liquids and 81 are solids at room temperature.

Substances may contain one element or several different elements. Hydrogen and oxygen are both elements. Water is made up of the two elements hydrogen and oxygen.

The elements may just be mixed together or they can be chemically combined (joined together). Air contains a **mixture** of gases that are not joined together. In water, the two elements are chemically joined together.

The table below shows some of the elements in some common substances.

Figure 2.1 *The elements in some common substances*

Substance	Elements	Type of substance
aluminium	aluminium	element
iron	iron	element
air	nitrogen, oxygen and argon	mixture
brass	copper and zinc	mixture
water	hydrogen and oxygen	compound
salt	sodium and chlorine	compound
sugar	carbon, hydrogen and oxygen	compound
methane	carbon and hydrogen	compound

Did you know?

The most common element in the Universe is hydrogen, which makes up 90% of all known matter. The most common element in the Earth's crust is oxygen, which makes up 46.6% of the crust by weight.

Elements are substances made from atoms which contain the same number of protons. Each element has its own individual set of chemical properties that make it react in certain ways. Each element has a fixed position in the **periodic table** (see section 5.2).

Symbols for elements

Each element is given an abbreviation of one or two letters called a symbol. For example, the symbol for hydrogen is H and the symbol for calcium is Ca.

Some elements have symbols based on Latin names. For example, the symbol for gold is Au. This symbol is derived from the Latin word *aurum*. The symbol for iron is Fe, derived from the Latin word *ferrum*.

Did you know?

Plumbers have their name derived from the Latin word plumbum, which means lead. This is because water pipes used to be made from lead and so the men that fitted and mended them were called plumbers.

The following table gives the names and symbols of some common elements.

Figure 2.2 *Some of the common elements and their symbols*

Element	Symbol	Element	Symbol
aluminium	Al	iron	Fe
argon	Ar	lead	Pb
barium	Ba	lithium	Li
beryllium	Be	magnesium	Mg
boron	B	manganese	Mn
bromine	Br	neon	Ne
calcium	Ca	nitrogen	N
carbon	C	oxygen	O
chlorine	Cl	phosphorus	P
copper	Cu	potassium	K
fluorine	F	silicon	Si
gold	Au	silver	Ag
helium	He	sodium	Na
hydrogen	H	sulphur	S
iodine	I	zinc	Zn

Did you know?

Every name tells a story. The element helium was discovered in the Sun before it was discovered on the Earth. It was named after the Greek word for the Sun, helios. Copper is named after the Latin word *cuprum*, which in turn was derived from the old name for Cyprus.

Compounds

Compounds are substances which contain two or more elements chemically joined together. Sodium chloride contains the elements sodium and chlorine chemically combined together. Carbon dioxide contains the elements carbon and oxygen chemically combined together.

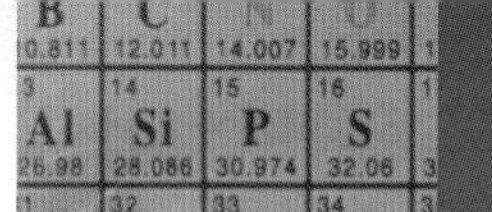

Most elements are found as compounds in their natural state. Iron in iron ore is combined with oxygen to give iron(III) oxide, aluminium in bauxite is combined with oxygen to give aluminium oxide. Only a few elements are found uncombined in their natural state. For example gold, copper and sulphur are found in the ground as gold, copper and sulphur.

Did you know?

The largest nugget of gold ever found was picked up in Australia in 1869. It weighed 70.92 kg and was named, for obvious reasons, *Welcome Stranger*. At today's prices it would be worth £400 000.

Naming compounds

The names of compounds are usually based on the elements that they are made from. The names of some elements change slightly when they are combined with other elements to form compounds. Usually the combined element has the ending *-ide*. The table below shows some of these names.

Name of element when on its own		Name when joined with another element
chlorine	→	chloride
bromine	→	bromide
iodine	→	iodide
nitrogen	→	nitride
oxygen	→	oxide
sulphur	→	sulphide

Rather than write the full name of a compound each time it is used, it is easier to use an abbreviated version called a formula. The formula of a substance shows:

- which elements have been combined together
- the ratio of the atoms of each of the elements that form it.

For example:

$MgCl_2$ represents 1 atom of magnesium joined with 2 atoms of chlorine
Na_2SO_4 represents 2 atoms of sodium, 1 atom of sulphur and 4 atoms of oxygen joined together.

Some atoms are found grouped together in many compounds. These groups are called radicals and are often formed from acids. For example, SO_4 represents the sulphate radical. It comes from sulphuric acid, H_2SO_4.

Writing the formula of a compound

There are many different compounds, and rather than learn each different formula it is possible to give each element or radical a valency or combining power which can then be used to write the correct formula. The combining power depends on the structure of the atoms making up the compound (see section 1.1).

Some elements can have more than one combining power. In these cases the roman numeral I, II or III in the brackets after their name shows which combining power the element is using in the compound.

Figure 2.3a *Valency or combining power of elements*

Combining power 1	Symbol	Combining power 2	Symbol	Combining power 3	Symbol
bromine	Br	barium	Ba	aluminium	Al
chlorine	Cl	calcium	Ca	iron(III)	Fe
hydrogen	H	copper	Cu	nitrogen	N
iodine	I	iron(II)	Fe		
lithium	Li	lead	Pb		
potassium	K	magnesium	Mg		
silver	Ag	oxygen	O		
sodium	Na	sulphur	S		
		zinc	Zn		

Figure 2.3b *Valency or combining power of radicals*

Combining power 1	Symbol	Combining power 2	Symbol	Combining power 3	Symbol
ammonium	NH_4	carbonate	CO_3	phosphate	PO_4
hydrogencarbonate	HCO_3	sulphate	SO_4		
hydrogensulphate	HSO_4				
hydroxide	OH				
nitrate	NO_3				
nitrite	NO_2				

The valency or combining power of the element or radical can be used to work out the formula of the substance using the following steps.

Step 1: From the name of the substance write down the symbols of the elements or radicals.
Step 2: Write the valency number above each symbol.
Step 3: Swap the valencies and write them below the other symbol.
Step 4: If the numbers are the same, then just write the symbols.
Step 5: Leave out any that are 1 and put brackets around the radicals if there are two or more.

Examples

Name of compound	Symbols of elements or radicals	Valencies	Formula of compound
Aluminium oxide	Al O	$Al^3\ O^2$	Al_2O_3
Zinc sulphate	Zn SO_4	$Zn^2\ SO_4{}^2$	$ZnSO_4$
Sodium sulphide	Na S	$Na^1\ S^2$	Na_2S
Potassium carbonate	K CO_3	$K^1\ CO_3{}^2$	K_2CO_3
Iron (III) sulphate	Fe SO_4	$Fe^3\ SO_4{}^2$	$Fe_2(SO_4)_3$
Ammonium sulphate	NH_4 SO_4	$NH_4{}^1\ SO_4{}^2$	$(NH_4)_2SO_4$

The use of prefixes in naming compounds

Some compounds contain the same elements combined in different positions. The number of atoms of each element can be shown using a prefix.

Figure 2.4 *Prefixes used for numbers*

Number	Prefix
1	mon
2	di
3	tri
4	tetra

The following compounds have prefixes in their names.

carbon monoxide	CO
carbon dioxide	CO_2
sulphur dioxide	SO_2
sulphur trioxide	SO_3
dinitrogen tetroxide	N_2O_4

Special cases

The correct formula of an element or compound can usually be worked out from its name but there are some special cases where this is not possible.

1 Gases

Most gaseous elements, except the noble gases, form molecules (groups of atoms) rather than remaining as atoms (see section 1.2). Their formulae must show this. For example:

Element	**Formula of molecule**
hydrogen	H_2
nitrogen	N_2
oxygen	O_2
chlorine	Cl_2
bromine	Br_2
ozone	O_3

The noble gases only have one atom:

Name of element	**Formula**
helium	He
argon	Ar

2 Acids

Name of acid	**Formula**
hydrochloric acid	HCl
sulphuric acid	H_2SO_4
nitric acid	HNO_3

3 Other common formulae

water	H_2O
ammonia	NH_3
methane	CH_4

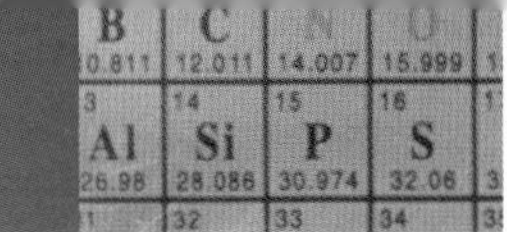

Representing chemical reactions

Many chemicals are harmful in some way or another. The diagrams below show some of the hazard symbols used to identify harmful chemicals.

Figure 2.5

Oxidising
These substances provide oxygen which allows other materials to burn more fiercely.

Highly flammable
These substances easily catch fire.

Toxic
These substances can cause death. They may have their effects when swallowed or breathed in or absorbed through the skin.

Harmful
These substances are similar to toxic substances but less dangerous.

Corrosive
These substances attack and destroy living tissues, including eyes and skin.

Irritant
These substances are not corrosive but can cause reddening or blistering of the skin.

Chemical change

A chemical change is called a chemical reaction. In a reaction, the substances which react together are called **reactants** and the new substances formed are called **products**. Magnesium reacts with hydrochloric acid to produce hydrogen and magnesium chloride. The reactants are magnesium and hydrochloric acid. The products are hydrogen and magnesium chloride. A chemical change is one in which the products have different chemical properties from the reactants.

The following observations could be an indication that a chemical change is taking place:

- bubbles of gas are formed
- a colour change is seen
- the substances get hotter or colder
- the original substance disappears
- a new substance appears
- the substances burn or glow
- a **precipitate** (an insoluble substance in suspension) forms.

Chemical equations

Chemical equations are a way of describing exactly what happens in a reaction. A word equation only includes the names of the reactants and products. A chemical equation is an abbreviated version of a word equation where the names of the reactants and products are replaced by their formulae.

Chemical equations must balance – there must be the same number of atoms of each element on each side of the equation. An equation does not represent a chemical reaction unless it is completely balanced.

The examples below illustrate the stages involved in writing chemical equations.

Reaction 1: Magnesium ribbon burns in air with a brilliant light to form magnesium oxide.

The magnesium reacts with the oxygen in the air so the equation will only include magnesium and oxygen.

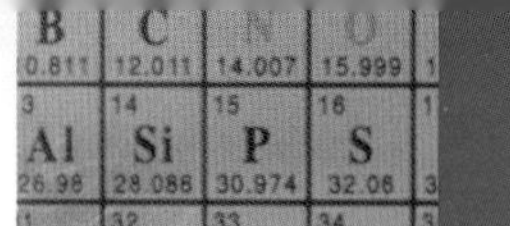

Word equation: magnesium + oxygen → magnesium oxide

The word equation is first re-written as an unbalanced formula equation

$$Mg + O_2 \rightarrow MgO$$

Note: O_2 is used because oxygen is a molecule made up of two oxygen atoms.

In terms of the atoms: 1'Mg' + 2 'O' → 1 'Mg' + 1 'O'

The numbers of oxygen atoms are not equal on each side. The equation must be balanced by putting numbers in front of each formula. The formulae for each reactant or product must never be changed. It may be helpful to put boxes in front of each formula and then only put numbers in these boxes. So in order to make the number of atoms for each element the same on each side,

$$\square Mg + \square O_2 \rightarrow \square MgO$$

$$\boxed{2} Mg + \boxed{1} O_2 \rightarrow \boxed{2} MgO$$

$$2Mg + O_2 \rightarrow 2MgO$$

In terms of the atoms there are: 2 'Mg' + 2 'O' → 2 'Mg' + 2'O' so the equation is balanced.

Reaction 2: Sulphur burns in oxygen with a blue flame to give sulphur dioxide.

Word equation: sulphur + oxygen → sulphur dioxide

Formula equation: $S + O_2 \rightarrow SO_2$

In terms of the atoms: 1 'S' + 2 'O' → 1 'S' and 2'O'

The equation is already balanced and so the balanced chemical equation is:

$$S + O_2 \rightarrow SO_2$$

Reaction 3: Zinc reacts with hydrochloric acid to give zinc chloride and hydrogen.

Word equation: zinc + hydrochloric acid → zinc chloride + hydrogen

Formula equation: $Zn + HCl \rightarrow ZnCl_2 + H_2$

In terms of the atoms: 1'Zn' + 1 'H' + 1 'Cl' → 1 'Zn' + 2 'Cl' + 2 'H'

The equation is not balanced. The numbers of chlorine and hydrogen atoms are not the same on each side.

$$\square Zn + \square HCl \rightarrow \square ZnCl_2 + \square H_2$$

$$\boxed{1} Zn + \boxed{2} HCl \rightarrow \boxed{1} ZnCl_2 + \boxed{1} H_2$$

$$Zn + 2HCl \rightarrow ZnCl_2 + H_2$$

In terms of the atoms: 'Zn' + 2 'H' + 2 'Cl' → 'Zn' + 2 'H' + 2 'Cl'

The balanced chemical equation is therefore:

$$Zn + 2HCl \rightarrow ZnCl_2 + H_2$$

Note: 2HCl means that there are two hydrogen atoms *and* two chlorine atoms. The number in front of the formula applies to all the atoms in a compound not just the first element.

Use of state symbols in equations

The states of the reactants and products can be shown in brackets after each formula:

(s) means solid
(l) means liquid
(g) means gas
(aq) means a solution in water (an aqueous solution)

Examples of the use of state symbols are:

$$2Mg(s) + O_2(g) \rightarrow 2MgO(s)$$

$$Zn(s) + 2HCl(aq) \rightarrow ZnCl_2(aq) + H_2(g)$$

Ionic equations

Special equations, called ionic equations, are used to show the **ions** taking part in a reaction (see section 1.2). These are often reactions involving **precipitation**, **neutralisation** and electrolysis.

Reaction 1: When a solution of silver nitrate is added to a solution of magnesium chloride a white precipitate of silver chloride is formed in a solution of magnesium nitrate.

The chemical equation for the reaction is:

$$2AgNO_3(aq) + MgCl_2(aq) \rightarrow 2AgCl(s) + Mg(NO_3)_2(aq)$$

The equation can be re-written in terms of the ions present and their states as:

$$2Ag^+(aq) + 2NO_3^-(aq) + Mg^{2+}(aq) + 2Cl^-(aq) \rightarrow 2AgCl(s) + Mg^{2+}(aq) + 2NO_3^-(aq)$$

Some of the ions, Mg^{2+} and NO_3^-, appear on both sides of the equation and are not changed during the course of the reaction. These ions are called spectator ions and can be left out of an ionic equation. So the ions remaining are:

$$2Ag^+(aq) + 2Cl^-(aq) \rightarrow 2AgCl(s)$$

which simplifies to:

$$Ag^+(aq) + Cl^-(aq) \rightarrow AgCl(s)$$

An ionic equation must balance in terms of the charge on each side as well as the number of atoms/ions of each element on each side. The net charge on each side must be checked.

Reaction 2: Iron filings will react with copper sulphate solution to give solid copper in a solution of iron(II) sulphate.

The equation for the reaction is:

$$Fe(s) + CuSO_4(aq) \rightarrow Cu(s) + FeSO_4(aq)$$

The equation can be re-written in terms of the ions present and their states as:

$$Fe(s) + Cu^{2+}(aq) + SO_4^{2-}(aq) \rightarrow Cu(s) + Fe^{2+}(aq) + SO_4^{2-}(aq)$$

The spectator ion, SO_4^{2-}, can be left out of the ionic equation. So the ions remaining are:

$$Fe(s) + Cu^{2+}(aq) \rightarrow Cu(s) + Fe^{2+}(aq)$$

The ionic equation is balanced for both charge and atoms/ions of each element.

Topic questions

1 Work out the number of atoms of each element present from the formulae of the following compounds:

$NaCl$ CaO $MgBr_2$ $NaNO_3$ $ZnSO_4$ Al_2O_3 NH_4OH $Fe_2(SO_4)_3$

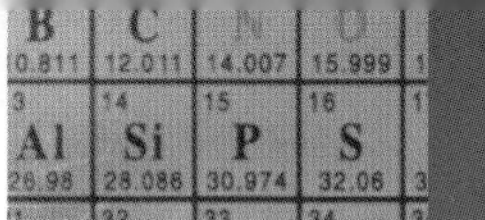

2 Name each of the following compounds:

a) KCl b) $ZnCl_2$ c) MgO d) KNO_3
e) $CuSO_4$ f) $Pb(NO_3)_2$ g) $AgNO_3$
h) Na_2SO_4 i) CH_4 j) CO_2

3 Write the formula of the following compounds:

a) silver chloride b) calcium sulphate
c) barium chloride d) sodium carbonate
e) copper nitrate f) aluminium nitrate
g) iron(III) oxide h) magnesium nitride
i) calcium phosphate j) ammonium sulphate

4 Balance the following equations Remember that **the formulae must not be altered.**

a) $Zn + O_2 \rightarrow ZnO$
b) $Mg + HCl \rightarrow MgCl_2 + H_2$
c) $KOH + H_2SO_4 \rightarrow K_2SO_4 + H_2O$
d) $Na_2CO_3 + HNO_3 \rightarrow NaNO_3 + H_2O + CO_2$
e) $FeSO_4 + NaOH \rightarrow Fe(OH)_2 + Na_2SO_4$

5 Write balanced chemical equations for the following word equations:

a) copper + sulphur → copper sulphide
b) carbon + oxygen → carbon monoxide
c) magnesium + nitric acid → magnesium nitrate + hydrogen
d) calcium carbonate + hydrochloric acid → calcium chloride + water + carbon dioxide
e) sodium hydroxide + sulphuric acid → sodium sulphate + water

6 Write balanced chemical equations for the following word equations:

a) iron + oxygen → iron(III) oxide
b) ammonia + sulphuric acid → ammonium sulphate
c) aluminium sulphate + sodium hydroxide → aluminium hydroxide + sodium sulphate
d) magnesium + nitrogen → magnesium nitride

7 Rewrite the following equations showing all the ions present and then write them as ionic equations without the spectator ions. Assume that the solids do not form ions. Show all state symbols.

a) $NaCl(aq) + AgNO_3(aq) \rightarrow AgCl(s) + NaNO_3(aq)$
b) $Mg(s) + CuSO_4(aq) \rightarrow Cu(s) + MgSO_4(aq)$
c) $ZnCl_2(aq) + 2AgNO_3(aq) \rightarrow 2AgCl(s) + Zn(NO_3)_2(aq)$
d) $Na_2SO_4(aq) + BaCl_2(aq) \rightarrow BaSO_4(s) + 2NaCl(aq)$
e) $Cu(NO_3)_2(aq) + 2NaOH(aq) \rightarrow Cu(OH)_2(s) + 2NaNO_3(aq)$

Summary

- All matter is made up from a limited range of simple substances called **elements**.
- Elements are substances made from atoms that contain the same number of protons.
- A **compound** is a substance that contains two or more elements chemically joined together.
- The formula of a substance uses chemical symbols to show which elements have been combined together in a compound.
- A **mixture** contains two or more different substances that are usually easy to separate.
- A chemical change is one in which the products have different chemical properties from the reactants.
- A chemical change is called a **reaction**.
- In a reaction the substances which react together are called **reactants** and the new substances formed are called **products**.

- Chemical equations are a way of describing exactly what happens in a reaction.
- A word equation only includes the names of the reactants and products and no description of the reaction itself.
- A chemical equation is an abbreviated version of a word equation in which the words are replaced by formulae.
- **Ionic equations** are used to show the ions taking part in a reaction.

2.2	
Co-ordinated	Modular
Introduction	07/08 (10.10)

Types of chemical reactions

Some common general reactions are given special names to describe the processes taking place.

Thermal decomposition reactions

This happens when a substance is split up into other substances by heat.

Reaction 1: When copper carbonate is heated it thermally decomposes to give copper oxide and carbon dioxide gas. The reaction can be represented as:

word equation: copper(II) carbonate → copper(II) oxide + carbon dioxide

chemical equation: $CuCO_3(s) \rightarrow CuO(s) + CO_2(g)$

Reaction 2: When limestone (calcium carbonate) is heated strongly it thermally decomposes to give calcium oxide and carbon dioxide. The reaction can be represented as:

word equation: calcium carbonate → calcium oxide + carbon dioxide

chemical equation: $CaCO_3(s) \rightarrow CaO(s) + CO_2(g)$

Thermal decomposition is an **endothermic reaction** (see sections 2.3 and 6.1).

Did you know?

Some substances are so thermally unstable that they will decompose explosively at room temperature. Stores of ammonium nitrate have been known to suddenly explode leaving behind a large crater where the buildings were.

Neutralisation reactions

Neutralisation is a reaction between an acid and a base (see section 5.6). A base is a metal oxide or hydroxide. If a base is soluble in water, it is called an alkali. Usually a base is added to an acid in order to neutralise it. Universal indicator can be used to show when all the acid has been neutralised. At that point the pH will equal 7 and the universal indicator will be green.

Reaction 1: When sodium hydroxide solution is added to hydrochloric acid it will neutralise it to form sodium chloride and water. The reaction can be represented as:

word equation: sodium hydroxide + hydrochloric acid → sodium chloride + water

chemical equation: $NaOH(aq) + HCl(aq) \rightarrow NaCl(aq) + H_2O(l)$

Reaction 2: When calcium hydroxide solution is added to nitric acid it will neutralise it to form calcium nitrate and water. The reaction can be represented as:

word equation: calcium hydroxide + nitric acid → calcium nitrate + water

chemical equation: $Ca(OH)_2(aq) + 2HNO_3(aq) \rightarrow Ca(NO_3)_2(aq) + 2H_2O(l)$

Did you know?

Acids and alkalis are used as weapons in the insect world. A bee stings by injecting an acid and an ant 'bite' is caused by a squirt of acid on the skin. Bee stings and ant bites can be treated with a mild alkali to neutralise the acid. A wasp sting is alkaline and can be treated with a mild acid to neutralise the alkali.

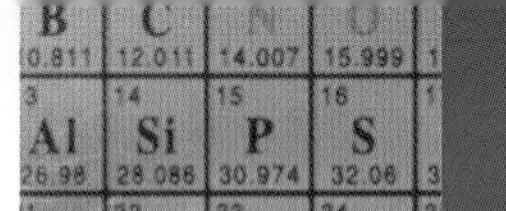

Displacement reactions

Displacement reactions happen when one element in a substance is replaced or 'pushed out' by another element. Hydrogen can be displaced from acids by metals and a metal can be displaced from a salt by a more reactive metal.

Reaction 1: Magnesium will react with sulphuric acid to form magnesium sulphate and hydrogen gas. The hydrogen in the acid is displaced by magnesium to form a salt called magnesium sulphate.

This displacement reaction can be represented as:

word equation: magnesium + sulphuric acid → magnesium sulphate + hydrogen

chemical equation: $Mg(s) + H_2SO_4(aq) \rightarrow MgSO_4(aq) + H_2(g)$

Reaction 2: Zinc will react with copper(II) sulphate in solution to form zinc sulphate and copper. The copper in copper(II) sulphate has been displaced by the more reactive zinc to form the salt zinc sulphate. This displacement reaction can be represented as:

word equation: zinc + copper(II) sulphate → copper + zinc sulphate

chemical equation: $Zn(s) + CuSO_4(aq) \rightarrow Cu(s) + ZnSO_4(aq)$

The ionic equation is: $Zn(s) + Cu^{2+}(aq) \rightarrow Cu(s) + Zn^{2+}(aq)$

Precipitation

Precipitation is a reaction between two soluble substances in solution that results in an insoluble product. The insoluble product appears as a suspension or precipitate.

In the reaction between solutions of silver nitrate and sodium chloride, silver chloride and sodium nitrate are produced. The silver chloride is insoluble and will be seen as a white precipitate. The sodium nitrate is a colourless solution. The reaction can be represented as:

word equation: silver nitrate + sodium chloride → silver chloride + sodium nitrate

chemical equation: $AgNO_3(aq) + NaCl(aq) \rightarrow AgCl(s) + NaNO_3(aq)$

The ionic equation is: $Ag^+(aq) + Cl^-(aq) \rightarrow AgCl(s)$

Oxidation and reduction

Oxidation occurs when a substance joins with oxygen or loses electrons. **Reduction** occurs when a substance loses oxygen or gains electrons.

Oxidation

A substance is oxidised if oxygen is added to it or it has gained more oxygen.

Reaction: When magnesium burns in air it forms magnesium oxide. The oxygen in the air has combined with the magnesium. The oxygen has oxidised the magnesium to magnesium oxide. The reaction can be represented as:

word equation: magnesium + oxygen → magnesium oxide

chemical equation: $2Mg(s) + O_2(g) \rightarrow 2MgO(s)$

Reduction

A substance, usually an oxide, is reduced when oxygen is taken away from it.

Reaction: When hydrogen is passed over heated copper(II) oxide, copper and water are formed. The copper(II) oxide has been reduced to copper by losing its oxygen.

The reaction can be represented as:

word equation: copper(II) oxide + hydrogen → copper + water

chemical equation: $CuO(s) + H_2(g) \rightarrow Cu(s) + H_2O(l)$

Oxidation and reduction will always occur together. When one substance is being oxidised another substance is being reduced. In the reaction between copper oxide and hydrogen

- the copper oxide has lost oxygen so has been reduced
- the hydrogen has gained oxygen so has been oxidised
- the hydrogen causes the reduction so is called a reducing agent
- the copper oxide provides the oxygen so is called the oxidising agent.

Reduction and oxidation always occur together in reactions and so these are often called redox reactions from REDuction and OXidation.

The gaining or losing of electrons

When copper metal is oxidised, the copper metal becomes copper ions. For this to happen, the copper metal has to lose electrons to a suitable electron acceptor (in this case oxygen).

$$Cu \rightarrow Cu^{2+} + 2e^-$$

So a broader definition of oxidation is, a substance is oxidised if it loses electrons during a reaction. Transfer of electrons will result in atoms turning into ions, ions turning into atoms, or ions changing their charge.

When a sodium atom loses an electron it forms a positive sodium ion.

$$Na - e^- \rightarrow Na^+$$

Sodium has been oxidised because it has lost an electron.

When a chloride ion loses an electron it forms a chlorine atom.

$$Cl^- - e^- \rightarrow Cl$$

The chloride ion has been oxidised because it has lost an electron.

When copper oxide is reduced by hydrogen, the copper ion becomes copper metal. For this to happen, the copper ion has to gain electrons from a suitable electron donor (in this case hydrogen).

$$Cu^{2+} + 2e^- \rightarrow Cu$$

So a broader definition of reduction is, a substance is reduced if it gains electrons during a reaction.

When a bromine atom gains an electron it forms a negative bromide ion. Bromine has been reduced because it has gained one electron.

$$Br + e^- \rightarrow Br^-$$

When a copper 2+ ion gains two electrons, a copper atom is formed. The copper 2+ ion has been reduced by gaining two electrons.

$$Cu^{2+} + 2e^- \rightarrow Cu$$

Oxidation and reduction will always occur together. When one substance is being oxidised another substance is being reduced. The electrons will be transferred from the substance being oxidised to the substance being reduced.

The ionic version of the reaction of magnesium with oxygen is

$$2Mg(s) + O_2(g) \rightarrow 2Mg^{2+}2O^{2-}(s)$$

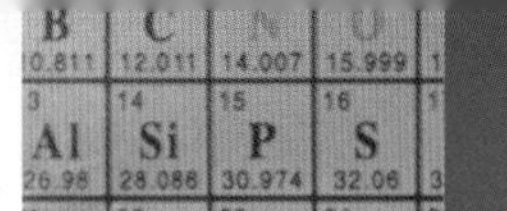

Each magnesium atom loses two electrons to form a magnesium 2+ ion.

$$Mg - 2e^- \rightarrow Mg^{2+}$$

The magnesium atom has been oxidised to a magnesium ion.

Each oxygen atom gains two electrons to form a negative oxide ion.

$$O + 2e^- \rightarrow O^{2-}$$

Oxygen has been reduced because it has gained two electrons.

Overall, the electrons are transferred from the magnesium atom to the oxygen atom forming new ions.

Reduction and oxidation have taken place in the reaction between magnesium and oxygen. So this reaction is another example of a redox reaction.

Reversible reactions

In most reactions the reactants will change into products and the reaction will stop. In a **reversible reaction** the products can turn back into reactants. Usually a balance is set up between the reactants and products and both are present at the same time. Reversible reactions are also called equilibrium reactions.

When nitrogen and hydrogen react together in the Haber Process (see section 3.2) they form ammonia. The ammonia can then decompose to form nitrogen and hydrogen again.

$$\text{nitrogen} + \text{hydrogen} \rightleftharpoons \text{ammonia}$$
$$N_2(g) + 3H_2(g) \rightleftharpoons 2NH_3(g)$$

The 'double-headed' arrow, $\rightleftharpoons$, shows that the reaction is reversible.

Topic questions

1 For each of the reactions, state whether it is a:

thermal decomposition reaction, neutralisation reaction, displacement reaction, precipitation reaction, reversible reaction, oxidation reaction or reduction reaction.

Give a reason for your choice.

a) When magnesium is burned in oxygen magnesium oxide is formed.
b) Insoluble barium sulphate is produced when a solution of barium nitrate is added to a solution of sodium sulphate.
c) When copper carbonate is heated it splits up into copper oxide and carbon dioxide.
d) Dilute hydrochloric acid reacts with a solution of sodium hydroxide to produce sodium chloride and water.
e) The heating of mercury oxide produces mercury and oxygen.
f) If blue copper(II) sulphate crystals are heated water is produced and the crystals go white. If water is added to the white copper(II) sulphate heat is produced and the copper(II) sulphate goes blue.
g) An iron nail put into a solution of copper(II) sulphate becomes coated in a brown layer of copper.

2 For each of the equations below state whether it is an oxidation or a reduction reaction. Give a reason for your choice. (State symbols have been omitted.)

a) $C + O_2 \rightarrow CO_2$
b) $2Ag_2O \rightarrow 4Ag + O_2$
c) $Cu - 2e^- \rightarrow Cu^{2+}$
d) $Pb^{2+} + 2e^- \rightarrow Pb$
e) $Fe^{2+} - e^- \rightarrow Fe^{3+}$
f) $Br_2 + 2e^- \rightarrow 2Br^-$
g) $K - e^- \rightarrow K^+$

3 Identify the oxidation and reduction reactions in each of the following reactions. Give the reasons for your choices in each case. (State symbols have been omitted.)

a) $PbO + H_2 \rightarrow Pb + H_2O$
b) $CuO + CO \rightarrow Cu + CO_2$
c) $2FeO + C \rightarrow 2Fe + CO_2$
d) $Zn + Fe^{2+} \rightarrow Fe + Zn^{2+}$
e) $Mg + Cu^{2+} \rightarrow Cu + Mg^{2+}$

Summary

- **Thermal decomposition** occurs when a substance is split up into other substances by heat.
- **Neutralisation** is a reaction between an acid and a base to form a neutral solution.
- **Displacement** occurs when one element in a substance is replaced or 'pushed out' by another element.
- **Precipitation** is a reaction between two solutions that results in an insoluble product.
- **Oxidation** is the gain of oxygen by a substance.
- A substance is oxidised if oxygen is added to it or it has gained more oxygen.
- A substance is oxidised if it loses electrons during a reaction.
- **Reduction** is the loss of oxygen from a substance.
- A substance, usually an oxide, is reduced when oxygen is taken away from it.
- A substance is reduced if it gains electrons during a reaction.
- In a **reversible reaction**, the products can turn back into reactants.

2.3	
Co-ordinated	Modular
Introduction	07/08 (10.10)

Exothermic and endothermic reactions

Exothermic reactions

Exothermic reactions *give out* heat. All combustion reactions are exothermic.

When methane is burnt in air, carbon dioxide and water are formed together with a large amount of heat. The reaction can be represented as:

word equation: methane + oxygen → carbon dioxide + water

chemical equation: $CH_4(g) + 2O_2(g) \rightarrow CO_2(g) + 2H_2O(l)$

Endothermic reactions

Endothermic reactions *take in* heat and get colder, or need heat to make them happen. When solutions of magnesium chloride and sodium carbonate are added together a reaction takes place to form magnesium carbonate and sodium chloride. As the products are formed the temperature of the mixture falls and is colder than that of either of the original solutions. The reaction can be represented as:

word equation: magnesium chloride + sodium carbonate → magnesium carbonate + sodium chloride

chemical equation: $MgCl_2(aq) + Na_2CO_3(aq) \rightarrow MgCO_3(aq) + 2NaCl(aq)$

There is more information about exothermic and endothermic reactions in section 6.1.

Summary

- **Exothermic reactions** are reactions which give out heat and get warmer.
- **Endothermic reactions** are reactions which take in heat and get colder.

Examination questions

1 Liquid ammonia is toxic.
Road tankers carrying liquid ammonia must display a hazard warning symbol.

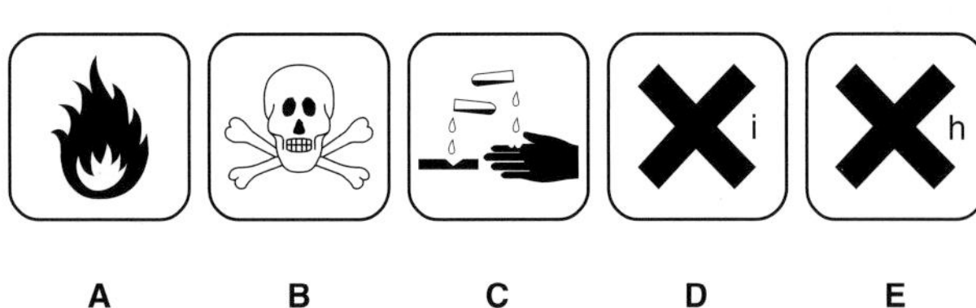

Which hazard warning symbol, A to E, should be displayed on the tanker? *(1 mark)*

2 a) A student studied the effect of temperature on the rate of reaction between hydrochloric acid and sodium thiosulphate.

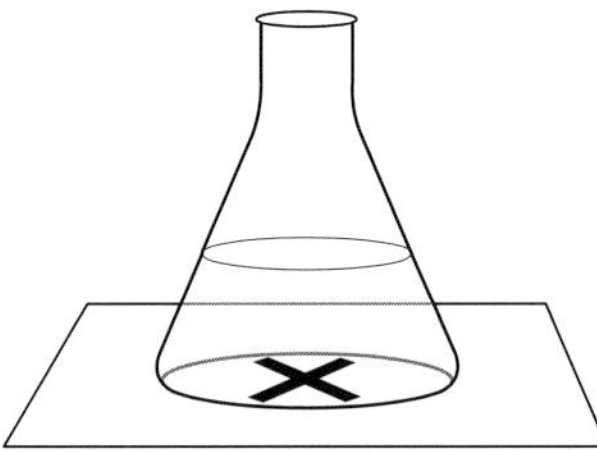

- The student mixed 50 cm^3 of a sodium thiosulphate solution and 5 cm^3 of hydrochloric acid in a flask.
- The flask was placed over a cross.
- The student timed how long after mixing the cross could no longer be seen.

a) i) Balance the chemical equation for this reaction.

$$Na_2S_2O_3(aq) + HCl(aq) \rightarrow NaCl(aq) + H_2O(l) + SO_2(g) + S(s)$$

(1 mark)

ii) What causes the cross to be seen no longer? *(1 mark)*

b) The student then tried to make some magnesium sulphate. Excess magnesium was added to dilute sulphuric acid. During this reaction fizzing was observed due to the production of a gas.

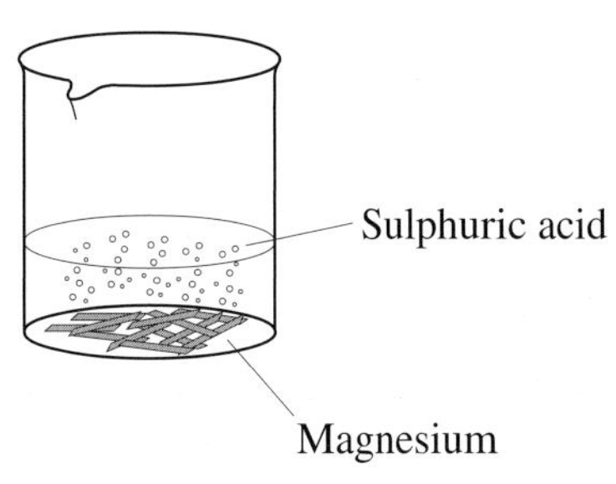

Complete and balance the chemical equation for this reaction.

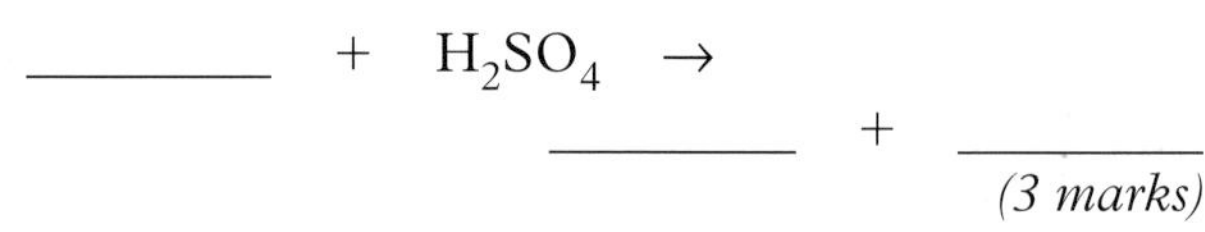

$$\_\_\_\_\_\_ + H_2SO_4 \rightarrow \_\_\_\_\_\_ + \_\_\_\_\_\_$$

(3 marks)

3 Bordeaux Mixture controls some fungal infections on plants. A student wanted to make some Bordeaux Mixture.

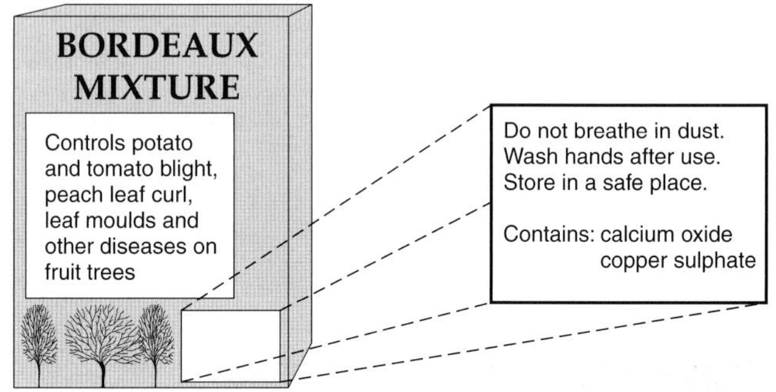

a) The student knew that calcium oxide could be made by heating limestone. Limestone contains calcium carbonate, $CaCO_3$.

i) Write the word equation for this reaction. *(1 mark)*

ii) What type of reaction is this? *(1 mark)*

b) The student knew that copper sulphate, $CuSO_4$, could be made by the following general reaction.

acid + base → salt + water

i) What type of reaction is this? *(1 mark)*

ii) The base used is copper oxide. Name and give the chemical formula of the acid used. *(2 marks)*

4 a) Iron is produced in a blast furnace by the following reaction.

$$Fe_2O_3 + 3CO \rightarrow 2Fe + 3CO_2$$

Both oxidation and reduction take place in this reaction. Explain how. *(2 marks)*

b) Hydrogen is used for the industrial production of ammonia. It is obtained from the reaction between methane and steam. The equation for this reaction is:

$$CH_4 + H_2O \rightarrow 3H_2 + CO$$

Explain how you can tell that this equation is balanced. *(2 marks)*

Chapter 3
The atmosphere

Key terms air • atmosphere • catalyst • denitrifying bacteria • endothermic • eutrophication • exothermic • fossilisation • global warming • greenhouse effect • Haber process • nitrifying bacteria • ozone layer • photosynthesis • respiration • reversible reaction • synthesis • ultraviolet radiation

3.1	
Co-ordinated	Modular
11.9	06 (10.11)

Changes to the atmosphere

Air

Air is a mixture of gases which for the last 200 million years has had a fairly constant composition. The amount of water vapour in the air varies depending on where you are. In the middle of the Sahara Desert there is very little water vapour; in a tropical rainforest the air contains a lot of water vapour.

Figure 3.1
The composition of air

Gas	Approximate %
nitrogen	80
oxygen	20
small amounts of various other gases, including carbon dioxide, water vapour and noble gases (e.g. argon)	

Did you know?

Although there is only 1% of argon in the air, that is actually quite a large amount. In a normal-sized school laboratory (10 m × 10 m × 3 m) there is about enough argon to fill the passenger compartment of an average sized car (3 m^3).

Oxygen supports combustion. This means that substances burn in oxygen. Air is 'dilute' oxygen, substances burn much better in pure oxygen.

A laboratory test for oxygen

Oxygen will cause a splint that is glowing red hot to rekindle and start burning. (Sometimes this may be accompanied by a slight pop – be careful not to confuse this with the test for hydrogen.)

Evolution of the atmosphere

The Earth's **atmosphere** has not always had the same composition. During the first billion years whilst the Earth was forming there was lots of volcanic activity. This

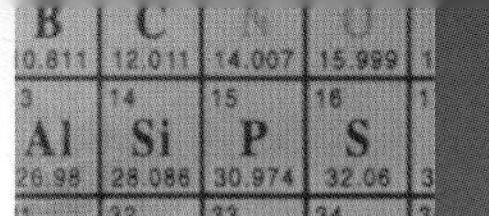

produced an atmosphere that was mainly made up of carbon dioxide, together with some ammonia (NH_3), methane (CH_4) and water vapour. There was little or no oxygen gas, so the Earth's atmosphere would have been similar to the present atmospheres of Mars and Venus. As the Earth gradually cooled the water vapour condensed to form seas and oceans.

What happened to the carbon dioxide?

Over millions of years the high carbon dioxide content gradually decreased. There were a number of reasons for this.

- Carbon dioxide reacted with sea water to form carbonates (mainly calcium carbonate) which was used to make the shells of many sea creatures. When the creatures died, their shells formed sediments which eventually became rocks such as chalk and limestone ($CaCO_3$).
- Carbon dioxide was removed by green plants through the process of **photosynthesis**. This process locks carbon dioxide up as glucose and releases oxygen.
- Plants and tiny animals (for example bacteria) became trapped in sediments and decayed. The effect of pressure on the decaying organisms produced fossil fuels, oil and natural gas. **Fossilisation** is a slow process. Most fossil fuels are formed from organisms fossilised about 300 000 000 years ago.

What effect did the increasing amount of oxygen have?

As more and more oxygen was released by the process of photosynthesis, the ammonia reacted with the oxygen to produce nitrogen and water.

The methane also reacted with oxygen to produce carbon dioxide and water.

Many micro-organisms on the Earth at that time could not tolerate oxygen. To them the increasing amount of oxygen was a pollutant and they began to be killed.

Other changes to the atmosphere

Some of the earliest forms of life on Earth were bacteria, among which were **nitrifying bacteria** that could convert ammonia into nitrates and **denitrifying bacteria** that could break down some of the nitrogen-containing compounds to form nitrogen gas. The green plants absorbed some of the nitrates from the ground to produce plant protein. The action of the bacteria and the plants reduced the ammonia content of the air and replaced it with nitrogen.

The action of sunlight on the oxygen in the atmosphere turned some of the oxygen molecules produced by photosynthesis (O_2) into molecules of ozone (O_3). This ozone formed a layer, the **ozone layer** around the Earth at a height varying from about 19 to 48 km. This layer, which was formed many millions of years ago, protects life on Earth by reducing the full effects of the Sun's cancer-causing **ultraviolet radiation**.

What is happening to the level of carbon dioxide now?

- Some of the carbon dioxide locked up as carbonate rocks is sometimes released back into the atmosphere through the eruption of volcanoes.
- **Respiration** releases the carbon dioxide locked up in carbohydrates.

- Large amounts of carbon dioxide are released by the burning of fossil fuels. These amounts are so large that the level of carbon dioxide in the atmosphere, which for millions of years had been stable, is now increasing.
- Because of the increase in atmospheric carbon dioxide there is a increase in the rate at which carbon dioxide reacts with sea water to form insoluble carbonates (mainly calcium carbonate). These carbonates continue to be deposited as

sediments or they may form soluble hydrogencarbonates (mainly calcium and magnesium). However, even though there is an increase in the rate at which carbon dioxide is being removed from the atmosphere it is not rapid enough to absorb the additional carbon dioxide now being released into the atmosphere. Increased amounts of carbon dioxide in the atmosphere lead to an increase in the **greenhouse effect** and **global warming**.

Summary

- For about 200 million years our atmosphere has been made up of about (80%) nitrogen, (20%) oxygen, together with small amounts of other gases such as carbon dioxide, water vapour and noble gases.
- The very early atmosphere is likely to have contained mainly carbon dioxide and water vapour with some methane and ammonia due to volcanic activity.
- Green plants removed some of the carbon dioxide during photosynthesis.
- Some of the carbon dioxide was locked up as rocks or fossil fuels.
- Nitrogen was produced as oxygen reacted with ammonia and the action of denitrifying bacteria.
- Some oxygen became ozone to form the ozone layer.
- Some carbon dioxide is being released by the action of volcanoes.
- Large quantities of carbon dioxide are released by the burning of fossil fuels.
- The continued formation of carbonates and hydrogencarbonates in the sea is too slow a process to lock up the extra carbon dioxide being released.

Topic questions

1 In the box are the names of the main gases found in the atmosphere.

argon carbon dioxide nitrogen oxygen water vapour

Use these words to answer the following questions. (You may use each word once.)
a) Which gas makes up most of the atmosphere?
b) Which gas is a noble gas?
c) Which gas is used in photosynthesis?
d) Which gas is made by photosynthesis?
e) Of which gas are there varying amounts depending on the local conditions?

2 In the box are the names of the main gases found in the Earth's early atmosphere.

ammonia carbon dioxide methane water vapour

Use these words to answer the following questions. (You may use each word once, more than once or not at all.)
a) Which gas produced the oceans when it cooled down?
b) Which gas was converted into nitrogen?
c) Which gas enabled plants to make protein?
d) Which gas was used to make limestone rocks?

3 Describe why the amount of carbon dioxide in today's atmosphere is less than it was when the Earth was forming.

4 Why is the amount of carbon dioxide in the atmosphere now gradually increasing?

3.2 Useful products from the air

Co-ordinated	Modular
11.6	07 (10.8)

Ammonia

Ammonia is a compound of nitrogen and hydrogen with the formula NH_3. It is made commercially by the Haber process. In this process nitrogen from the air is reacted with hydrogen. The process is widely used to make nitrogenous fertilisers. In fact about 80% of the ammonia produced is used for this purpose. Ammonia is also used to make nitric acid.

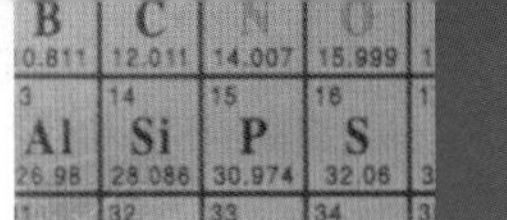

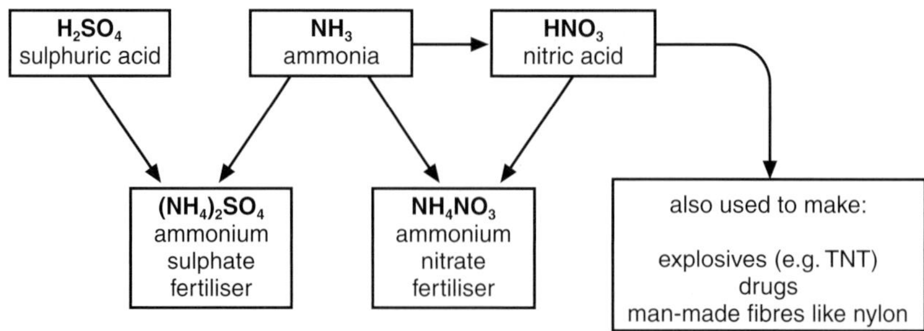

Figure 3.2
The uses of ammonia

Using ammonia to make nitric acid

Millions of tonnes of nitric acid are made each year, most of which is used in the manufacture of the fertiliser ammonium nitrate. Most nitric acid is made from ammonia. Three stages are involved:

Stage 1 A mixture of air and ammonia is passed over a platinum **catalyst** (see section 6.3) which is heated to a temperature of about 900°C. The oxygen in the air oxidises (see section 2.3) the ammonia into nitrogen monoxide (NO) and water.

Stage 2 The nitrogen monoxide gas is cooled and reacted with more oxygen to form nitrogen dioxide (NO_2).

Stage 3 Further reactions between nitrogen dioxide, oxygen and water produce nitric acid (HNO_3).

The Haber process

In the Haber process ammonia is made by the direct combination of nitrogen and hydrogen using iron as a catalyst to speed up the reaction.

$$\text{nitrogen} + \text{hydrogen} \rightleftharpoons \text{ammonia}$$
$$N_2(g) + 3H_2(g) \rightleftharpoons 2NH_3(g)$$

The raw materials for this process (nitrogen and hydrogen) are inexpensive because they are readily available. Nitrogen is in the air and hydrogen is in the compounds water (H_2O) and methane (natural gas) (CH_4).

Figure 3.3
An industrial plant producing ammonia by the Haber process

Did you know?

Almost all modern chemical factories are very expensive to build. To make the plant economically viable it has to be in operation for many years. Using readily-available raw materials greatly reduces the risk of the method being replaced by a cheaper one.

The nitrogen can be obtained from the air by burning natural gas in air to use up the oxygen. It is not the only way of getting nitrogen out of the air but it is a fairly inexpensive method.

The hydrogen can be obtained from natural gas (methane) or water. Both substances are relatively inexpensive. The method most frequently used is to react methane with steam in the presence of a suitable catalyst.

$$\text{methane} + \text{steam} \rightarrow \text{carbon dioxide} + \text{hydrogen}$$
$$CH_4(g) + 2H_2O(g) \rightarrow CO_2(g) + 4H_2(g)$$

In the Haber process the nitrogen and hydrogen are passed over a catalyst which is heated to about 450°C, at a pressure of about 200 atmospheres. Some of the hydrogen and nitrogen react to form ammonia. The mixture of gases leaving the catalyst is cooled. The ammonia liquifies and is removed. The unreacted nitrogen and hydrogen mixture is recycled.

The reaction between nitrogen and hydrogen to produce ammonia is a **reversible reaction** (see sections 2.3 and 6.2). This means that some of the ammonia can break down again into hydrogen and nitrogen depending on the temperature and pressure of the reaction.

Increasing the rates of reaction in the Haber process

The chemical industry is geared up to make new substances in a cost effective way. The conditions for any industrial chemical process are often chosen for financial rather than chemical reasons. Several of the above factors are made use of in the manufacture of ammonia by the Haber process.

The chemical industry prefers to produce new materials by a continuous process and at minimal cost. Reaction conditions are carefully selected to achieve both of these aims.

The reaction in which nitrogen and hydrogen are combined together to form ammonia is an exothermic, reversible process as the equation below shows.

$$N_2(g) + 3H_2(g) \rightleftharpoons 2NH_3(g) \quad (+\text{ heat transferred})$$

Industrial chemists make use of three of the effects already discussed in order to make this reaction as efficient as possible.

1. They alter the pressure of the reactants. (The reaction only involves gases.)
2. They select a suitable temperature for the reaction to take place.
3. They use a catalyst.

Did you know?

There is an important principle that can be applied to reversible reactions. It is called le Chatelier's principle and it states that if a reversible reaction is at equilibrium (i.e. the forward and reverse reactions are taking place at the same speed) then if a change is made to the conditions, the reaction responds in such a way that it tries to cancel out the effect of the change. In simple terms this means that if we increase the pressure, the reaction will respond by trying to lower the pressure and if we increase the temperature the reaction responds by trying to lower the temperature etc.

Changing the pressure

$$N_2(g) + 3H_2(g) \rightleftharpoons 2NH_3(g)$$

The reaction above shows that one molecule of nitrogen reacts with three molecules of hydrogen to make two molecules of ammonia. Since equal volumes of gases contain the same number of molecules (see section 1.3) it follows that one volume of nitrogen reacts with three volumes of hydrogen to make just two volumes of ammonia. This means that as the reaction proceeds there is a reduction in volume from four volumes (1 + 3) to just two volumes. Increasing the pressure means more ammonia is produced. Increased pressure also speeds up the reaction (see section 6.2). This is because compressing the gas increases its concentration. Figure 3.4 shows how the amount of ammonia produced changes as the pressure inside the reaction vessel changes.

In practice, a moderate pressure of between 150 and 300 atmospheres is chosen. Although high pressures give a higher yield, they are too expensive to maintain.

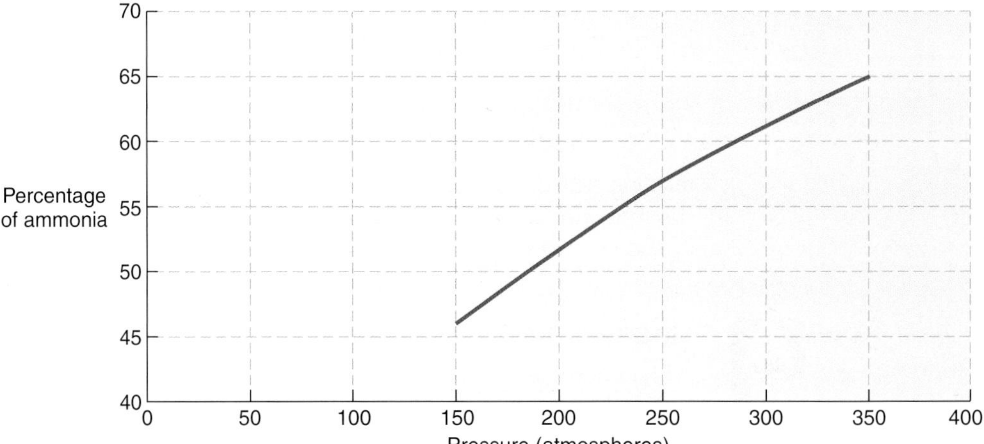

Figure 3.4 *The effect of pressure on the amount of ammonia produced*

Changing the temperature

$$N_2(g) + 3H_2(g) \rightleftharpoons 2NH_3(g) \quad (+ \text{ heat transferred})$$

The equation shows that the forward reaction is **exothermic**, that is, heat is produced (see section 6.2). We could write the equation in a simplified way:

nitrogen + hydrogen → ammonia (+ heat transferred)

If the reaction mixture is heated, it makes it harder for the reaction to give out heat. This means that less ammonia will be produced. So the cooler the reaction, the more ammonia is produced. The drawback here is that whilst the yield of ammonia will be quite high, the rate of the reaction at the low temperature will be very slow. Figure 3.5 shows how the yield of ammonia changes as the temperature changes.

Figure 3.5 shows that as the temperature is increased at different pressures, the amount of ammonia produced decreases.

Once again a compromise has to be made and temperatures between 400 and 500°C are usually used so that a reasonable yield of ammonia is produced reasonably quickly.

Using a catalyst

Some reactions can be speeded up by the use of a catalyst. The manufacture of ammonia is speeded up by the use of an iron catalyst.

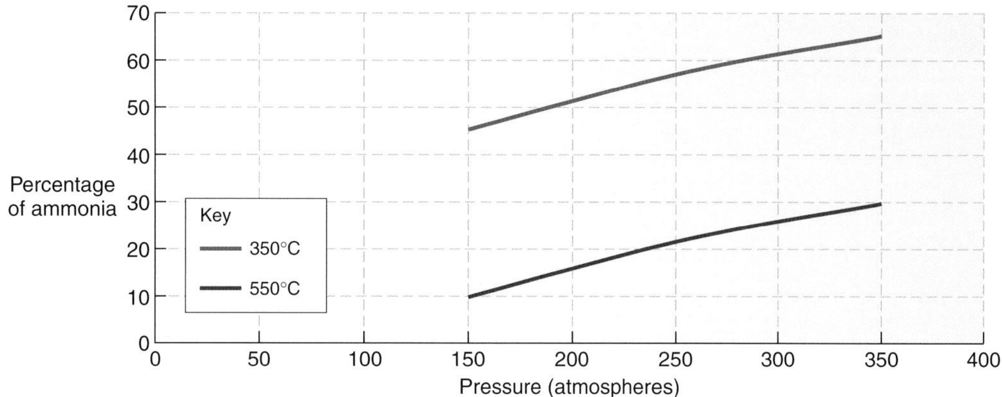

Figure 3.5 *The effect of temperature and pressure on the amount of ammonia produced*

Summary of the Haber process

In the **synthesis** of ammonia by the Haber process, nitrogen and hydrogen are combined together to form ammonia using a pressure of 150 to 300 atmospheres, a temperature between 400 and 500°C and an iron catalyst.

Nitrogenous fertilisers

Ammonia from the Haber process is used to produce the inexpensive nitrogenous (nitrogen-containing) fertilisers ammonium sulphate ($(NH_4)_2SO_4$) and ammonium nitrate (NH_4NO_3). Ammonium sulphate is made by the neutralisation of ammonia with sulphuric acid and ammonium nitrate is made by the neutralisation of ammonia with nitric acid.

Because these fertilisers are quite inexpensive there is a risk that they will not be used with sufficient care. If too much fertiliser is added to the land the plants cannot use all of it and some gets washed through the soil into rivers. This can cause the following environmental problems.

Figure 3.6 ▲ *Eutrophication in a river producing large amounts of algae*

1 Eutrophication of streams, rivers and lakes.

- Nitrogenous fertilisers encourage plant growth in water.
- Small plant-like organisms called algae grow rapidly.
- The small animals in the water which live on the algae cannot eat all the extra algae.
- Many of the algae die and are decomposed by bacteria.
- These bacteria take oxygen from the water.
- As the number of bacteria increases, the oxygen content of the water decreases and other larger organisms, such as fish, die of suffocation.

2 Nitrates in drinking water
Drinking water is taken from rivers. It is usually purified to kill harmful organisms but these processes do not remove any nitrates that have leached into the river from the land. If nitrates are present in drinking water, they can be converted in the body into chemicals called nitrosamines which are carcinogenic and may be a cause of some kinds of cancers of the alimentary canal.

Summary

- Ammonia is manufactured by the Haber process which requires temperatures of about 450°C and pressures of about 200 atmospheres.
- The reaction which uses the raw materials of nitrogen from the air and hydrogen from methane, is reversible.
- In order to maximise the yield of ammonia from the Haber process a number of economic factors associated with the reaction conditions need to be considered.
- Nitrogen is used to manufacture nitric acid, fertilisers and ammonia.
- Nitric acid can be made by the oxidation of ammonia.
- Nitrogenous fertilisers can cause environmental and health problems.

Topic questions

1 What is the formula of ammonia?

2 Answer the following questions about the manufacture of ammonia from nitrogen and hydrogen using the Haber process.
  a) From which source is the nitrogen obtained?
  b) Which two compounds are the source of hydrogen?
  c) What is the name of the catalyst used in the reaction?

3 a) Name the two main nitrogenous fertilisers.
  b) Give the chemical formulae of these substances.
  c) i) What is the name given to the polluting of rivers and streams by nitrogenous fertilisers?
    ii) Explain how nitrogenous fertilisers pollute rivers and streams.
  d) Why are nitrates in drinking water a health hazard?

4 In the Haber process for manufacturing ammonia gas from nitrogen and hydrogen, what are the advantages and disadvantages of using
  a) a very high pressure and
  b) a very high temperature?

Examination questions

1 Ammonia is a very important chemical.
  a) The table shows the percentage of ammonia used to make different substances.

Substances made from ammonia	Percentage (%) of ammonia used
fertilisers	75
nitric acid	10
nylon	5
others	10

  Copy the pie chart and shade the percentage of ammonia used to make nitric acid.

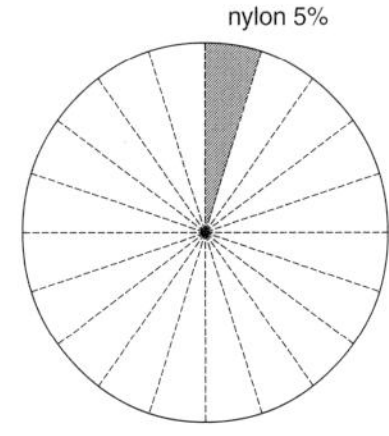

(1 mark)

  b) Ammonia gas is made by the reaction between nitrogen gas and hydrogen gas. Write a word equation to represent this reaction.

(1 mark)

  c) Nitrogen is one of the raw materials used to make ammonia. Nitrogen is obtained from air. This pie chart shows the proportion of nitrogen, oxygen and other gases in air. Label the area which represents the proportion of nitrogen in air.

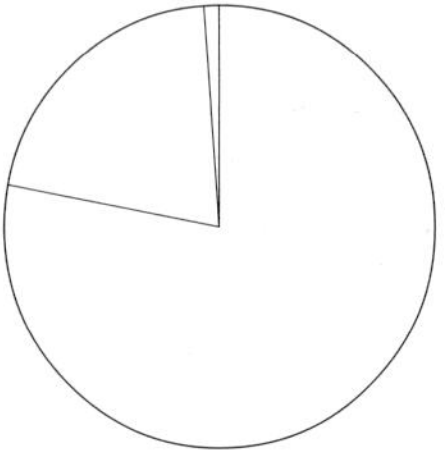

2 The air is a mixture of many gases. Some of these gases are shown in the table.

Name	Chemical formula
nitrogen	N_2
oxygen	O_2
argon	Ar
carbon dioxide	CO_2

a) Which of these gases are:
   i) elements
   ii) compounds? *(2 marks)*

b) Give *one* important use of nitrogen. *(1 mark)*

c) The amount of carbon dioxide in the air varies from place to place.
The amount of carbon dioxide in the countryside is often lower than in towns and cities. Explain why. *(2 marks)*

3 For 200 million years the proportions of the different gases in the atmosphere have been much the same as today. Over the past 150 years the amount of carbon dioxide in the atmosphere has increased from 0.03% to 0.04%.

a) Describe how carbon dioxide is released into the atmosphere by human and industrial activity. *(2 marks)*

b) Explain how the seas and oceans can decrease the amount of carbon dioxide in the atmosphere. *(3 marks)*

c) i) Give **one** reason why the amount of carbon dioxide in the atmosphere is increasing gradually. *(1 mark)*
   ii) Give **one** effect that increasing levels of carbon dioxide in the atmosphere may have on the environment. *(1 mark)*

4 The Haber process is used to make ammonia NH_3. The table shows the percentage yield of ammonia at different temperatures and pressures.

Pressure (atmospheres)	Percentage (%) yield of ammonia at 350°C	Percentage (%) yield of ammonia at 500°C
50	25	5
100	37	9
200	52	15
300	63	20
400	70	23
500	74	25

a) i) Use the data in the table to draw two graphs. Plot percentage (%) yield of ammonia on the vertical axis and pressures in atmospheres, on the horizontal axis. Draw one graph for a temperature of 350°C and the second graph for a temperature of 500°C. Label each graph with its temperature. *(4 marks)*
   ii) Use your graphs to find the temperature (°C) and pressure (atmospheres) needed to give a yield of 30% ammonia. *(1 mark)*
   iii) On the grid sketch the graph you would expect for a temperature of 450°C. *(1 mark)*

b) i) The equation represents the reaction in which ammonia is formed.

$$N_{2(g)} + 3H_{2(g)} \rightleftharpoons 2NH_{3(g)} + \text{heat}$$

What does the symbol $\rightleftharpoons$ in this equation tell you about the reaction? *(1 mark)*

   ii) Use your graphs and your knowledge of the Haber process to explain why a temperature of 450°C and a pressure of 200 atmospheres are used in industry. *(5 marks)*

Countryside – low amount of carbon dioxide

Towns and cities – high amount of carbon dioxide

Chapter 4
The Earth

Key terms

addition polymerisation • alkanes • alkenes • alloys • anode • bauxite • biodegradable • blast furnace • bromine water • carbon dioxide • cathode • coke • combustion • corrosion • cracking • crude oil • crust • cryolite • deposition • electrolysis • erosion • extrusive rock • flammable • fossil fuels • fractional distillation • galvanising • global warming • greenhouse effect • haematite • hydrocarbons • igneous rocks • intrusive rock • lava • limewater • magma • mantle • metamorphic rocks • minerals • monomers • non-renewable (finite) • ores • polymerisation • polymers • reactivity series • resources • rusting • sacrificial protection • saturated hydrocarbons • sedimentary rocks • smelting • structural formula • thermal decomposition • thermit process • transportation • unsaturated hydrocarbons • viscosity • weathering

4.1	
Co-ordinated	Modular
11.10	06 (10.2)

The rock record

Types of rock

Igneous rocks

Igneous rocks are formed when **magma** (the liquid rock in the Earth's **mantle**) cools and solidifies. If the magma solidifies beneath the Earth's surface it produces rocks with large crystals. These rocks are called **intrusive rocks**. Granite is an intrusive rock. If the magma erupts from the Earth's surface as **lava** it cools much more rapidly so the rocks formed have very small crystals. This type of rock is called **extrusive rock**. Basalt is an extrusive rock.

Sedimentary rocks

Sedimentary rocks are formed by the **deposition** of material on river bottoms and the sea bed. As the thickness of this sediment increases, the layers underneath become compressed and form sedimentary rocks. These rocks contain evidence of how and when they were formed:

- They may contain fossils.
- The different thicknesses and appearance of the layers show that there were periods of discontinuous deposition.
- There may be ripple marks in the rock, formed by waves in the water when the deposition occurred.
- Usually the youngest rocks are on top, but sometimes the layers may be tilted, folded, fractured (faulted) or even turned upside-down. This is because the Earth's crust is unstable and the rocks have been subjected to large forces.

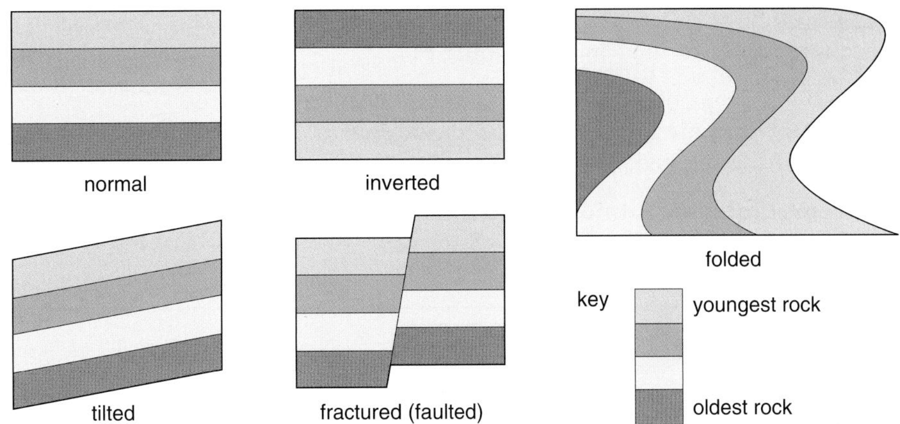

Figure 4.1 *Sedimentary rock structures*

Sandstone and mudstone are sedimentary rocks.

Limestone is also a sedimentary rock. It is formed by particles of calcium carbonate settling on the sea floor. Often the calcium carbonate is from the shells of dead sea creatures.

Metamorphic rocks

Metamorphic rocks are caused by movement in the Earth. This movement can create very high temperatures and pressures. Rocks subjected to these conditions may change to become metamorphic rocks.

Slate and marble are metamorphic rocks formed from mudstone and limestone, respectively. Schist is another example of a metamorphic rock.

Tectonic activity

The Earth's surface is made of a number of 'pieces' which fit together like a huge, spherical, jigsaw puzzle. These pieces are called **tectonic plates**. The plates are slowly moving. If the plates move towards each other they may buckle and produce mountain ranges (like the Himalayas). This process takes many millions of years. These new mountain ranges replace others formed in the distant past that have been worn down by weathering and erosion.

The collision between the plates causes the rocks to be put under great pressure. Sometimes the rocks are forced beneath the Earth by the pressure. This can cause the rocks to be heated to a very high temperature.

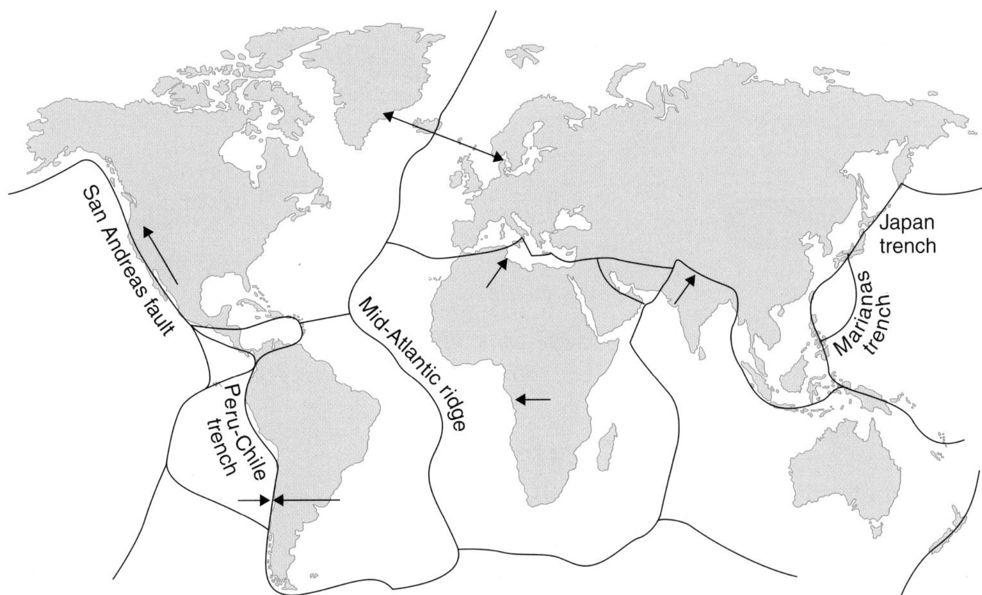

Figure 4.2 *The Earth, showing plate boundaries and direction of movement*

Summary

- **Igneous rocks** are formed by the cooling of molten rock.
- **Sedimentary rocks** are formed by eroded material settling on the sea bed.
- **Metamorphic rocks** are formed by the action of high temperatures and/or high pressures on other types of rock.
- The **rock cycle** is the movement of rock from the Earth's mantle to the surface and back again.
- Movement of the Earth's **tectonic plates** can cause rocks to tilt, fold or fracture (fault). It can also create the high temperatures and/or pressures needed to form metamorphic rocks.

Topic questions

1 Complete the following sentence by choosing the correct word from the alternatives given.

A volcano throws out **gas/lava**. This material cools **rapidly/slowly** to produce **extrusive/intrusive** rocks with **large/small** crystals. An example of this type of rock is **basalt/granite/slate**. If a large mass of underground **lava/magma** cools down the rock formed is **extrusive/intrusive**. **Basalt/granite/slate** formed this way.

2 Metamorphic rocks can be formed when other rocks get hot.

a) What might cause the rocks to get hot?
b) Other than heat, what else can produce metamorphic rocks?

3 How are mountain ranges formed?

4 Why are younger sedimentary rocks usually found on top of older sedimentary rocks?

4.2	
Co-ordinated	Modular
11.4	05 (10.5)

Useful products from metal ores

Rocks and minerals

Most rocks are not pure substances, they are mixtures of different **minerals**. Minerals are usually compounds but some – like gold and sulphur – are elements. Most minerals are unreactive; if they weren't, they would soon be weathered away. Some minerals can be made to react, and useful materials can be extracted from them. These materials are called **ores**.

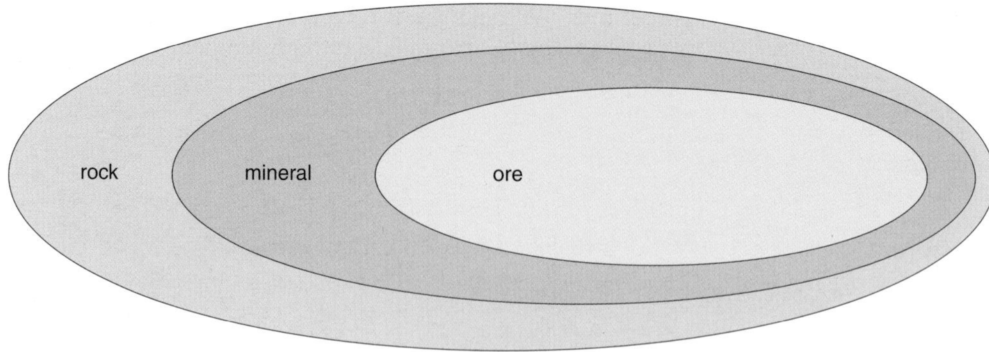

Figure 4.3 *A Venn diagram showing the relationship between rocks, minerals and ores*

Did you know?

The Earth's crust contains about 28% silicon and 8% aluminium but it is not possible to extract these elements economically from most of the minerals that contain them.

For a mineral to be classed as an ore it must be possible to extract useful amounts of material from it – and to do so economically.

Metal extraction

Most metals are extracted from ores. The exceptions are very unreactive metals like gold and silver which are found as the element. Most ores contain a metal oxide or a substance that can easily be changed into an oxide. To obtain a metal from its oxide the oxygen must be removed. This is an example of reduction. There are two main methods of removing the oxygen from the metal oxide. The method used depends on the reactivity of the metal. To decide which method is used it is helpful to include carbon in the **reactivity series** even though this is not a metal.

Figure 4.4
The reactivity series for metals showing the position of carbon

potassium
sodium
calcium
magnesium
aluminium
carbon – not a metal but included in the series
zinc
iron
tin
lead
hydrogen – not a metal but included in the series
copper
silver
gold
platinum

Any metal in the reactivity series can displace a metal lower in the series from its compounds. So carbon can displace the metals zinc, iron and copper from their compounds. This process of extracting metals from their ores by heating them with carbon is called **smelting**. Metals above carbon have to be extracted by another method. This process is called **electrolysis**.

Did you know?

The metals potassium, sodium, calcium, magnesium, strontium and barium were all discovered using electrolysis by Sir Humphry Davy between 1807 and 1808. Sir Humphry Davy also invented the coal miner's safety lamp, proved chlorine was an element, and was the first person to suggest that all acids contained the element hydrogen.

Uses of the reactivity series

Knowing the relative position of one metal to another in the reactivity series can be very helpful in predicting and explaining what happens in many chemical reactions.

- Metals high in the series tend to react or stay as a compound.
- Metals low in the series tend to stay as metals or become metals if they are part of a compound.

Predicting the reaction between a metal oxide and another metal

When a metal oxide is heated with a second metal it is possible for the second metal to remove the oxygen from the metal oxide. A metal higher in the reactivity series will always remove oxygen from the oxide of a metal that is lower than it in the series.

For example:

magnesium + copper(II) oxide → magnesium oxide + copper

$$Mg(s) + CuO(s) \rightarrow MgO(s) + Cu(s)$$

but

copper + magnesium oxide → NO REACTION

There is a competition between the two metals for the oxygen. The metal which is higher in the series will gain the oxygen and form an oxide, and the metal which is lower in the reactivity series will lose the oxygen and will form a metal.

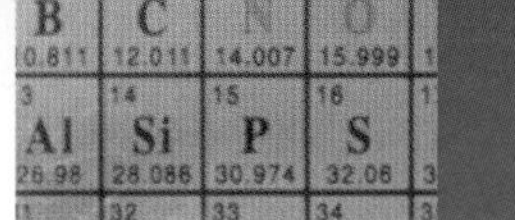

The Thermit process

The Thermit process is a method of joining two lengths of railway track together using the greater reactivity of aluminium compared to iron. A mixture of aluminium powder and iron(III) oxide in a crucible is positioned over the gap between the rails and ignited using a magnesium ribbon. A vigorous exothermic reaction takes place forming molten iron and aluminium oxide.

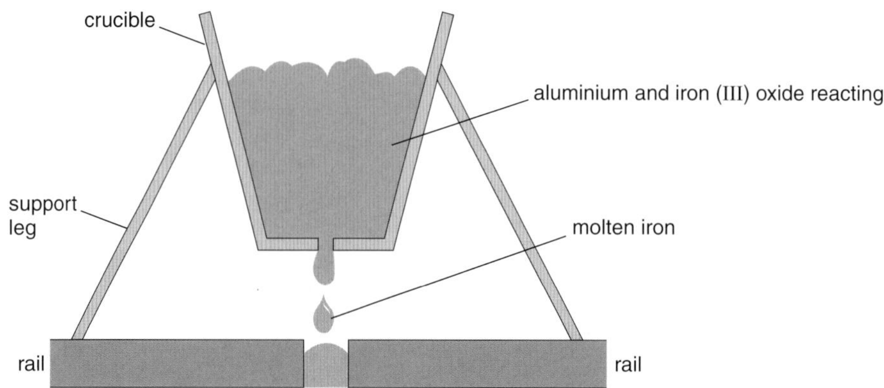

Figure 4.5 *The Thermit process for welding together two lengths of railway line*

The iron falls through the hole in the crucible and into the gap between the rails. When the iron cools and solidifies, the two lengths of rail will be joined.

Predicting the method of extraction of a metal from its ore

The position of a metal in the reactivity series can be used to predict the likely extraction process. Aluminium and iron are extracted from their oxides, Al_2O_3 and Fe_2O_3, respectively. Aluminium is high in the series and is a reactive metal. Aluminium is extracted from aluminium oxide by the high energy process of electrolysis. Iron is lower in the series and is much less reactive. Iron is extracted from iron(III) oxide by heating with carbon.

Iron

Iron is obtained from its ore by smelting. This process is carried out in a **blast furnace**.

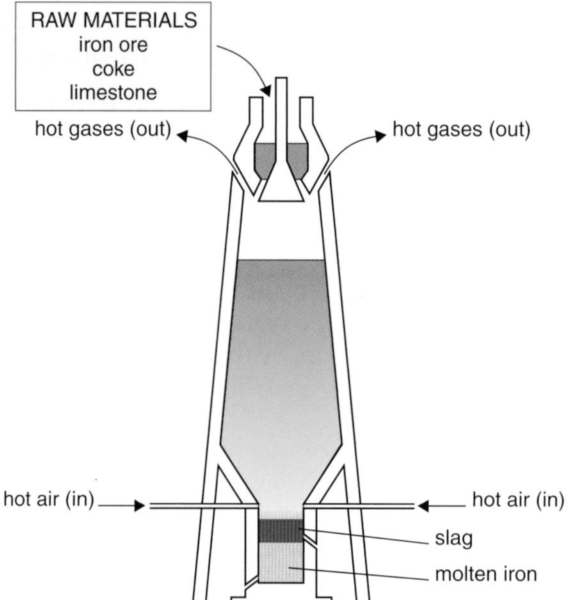

Figure 4.6 *Diagram of the inside of a blast furnace*

Iron ore (**haematite**, mainly iron(III) oxide, Fe_2O_3), **coke** (fairly pure form of carbon obtained by heating coal) and limestone are put into the top of the blast furnace. Hot air is blown in at the bottom.

In the blast furnace the following reactions take place:

1 Coke burns in the hot air blast producing **carbon dioxide**. (This reaction also helps to keep the furnace hot.)

carbon (coke) + oxygen → carbon dioxide

$$C(s) + O_2(g) \rightarrow CO_2(g)$$

2 The carbon dioxide produced is reduced (see Chapter 2) to carbon monoxide by some more coke.

carbon dioxide + carbon → carbon monoxide

$$CO_2(g) + C(s) \rightarrow 2CO(g)$$

3 The iron(III) oxide is reduced to iron by either carbon or carbon monoxide. The carbon (coke) or carbon monoxide is oxidised to CO_2.

iron(III) oxide + carbon → carbon dioxide + iron

$$2Fe_2O_3(s) + 3C(s) \rightarrow 3CO_2(g) + 4Fe(l)$$

iron(III) oxide + carbon monoxide → carbon dioxide + iron

$$Fe_2O_3(s) + 3CO(g) \rightarrow 3CO_2(g) + 2Fe(l)$$

Did you know?

Of the total mass of material produced by the blast furnace only about 25% is iron. About 66% is the gases which the reactions produce. Most of the gas is nitrogen from the hot air blast.

The temperature inside the blast furnace reaches 1400°C. This is hot enough for the iron to melt and the molten iron runs to the bottom of the furnace.

Iron ore is not very pure – it contains a lot of sand (silicon dioxide, SiO_2). Sand does not melt in the blast furnace. If it were left there it would become acidic impurities in the iron. The impurities are removed by the limestone added to the blast furnace.

sand + limestone → calcium silicate (slag) + carbon dioxide

$$SiO_2(s) + CaCO_3(s) \rightarrow CaSiO_3(l) + CO_2(g)$$

The calcium silicate (slag) melts in the blast furnace. It runs to the bottom and settles on top of the molten iron.

Uses of iron

Iron from the blast furnace is 'cast iron'. It contains about 4% carbon. This amount of carbon makes the iron very brittle. Iron can be converted into steel by passing oxygen through the molten cast iron. The oxygen reacts with some of the carbon to produce carbon dioxide. This lowers the amount of carbon in the iron. Other elements can be added to the iron to make **alloys** with special properties.

Figure 4.7 *Some alloys of iron and their uses*

Name of alloy	Composition	Properties	Uses
wrought iron	almost pure iron	easy to soften and shape	decorative iron work
mild steel	iron 99.5% carbon 0.5	quite hard but easy to work	buildings, car bodies
hard steel	iron 99% carbon 1%	very hard	ball bearings, blades for cutting tools
'stainless' steel	iron about 74% chromium 18% nickel 8%	resistant to corrosion	containers for corrosive substances, kitchenware
high speed steel	iron about 75% tungsten 18% chromium 4% vanadium 1% carbon 1%	very hard; not easily softened by high temperatures	cutting tools for metal working lathes

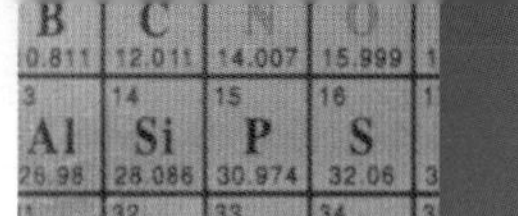

The rusting of iron

Like most transition metals, iron will **corrode** in the presense of air (oxygen) and water. In fact iron corrodes more quickly than most other transition elements. The corrosion of iron is called **rusting.** Rust is hydrated iron(III) oxide although the exact formula of rust is difficult to give.

Rusting as an oxidation reaction

Rusting is the oxidation of iron – the iron gains oxygen as part of the rusting reaction. If rust is assumed to be iron(III) oxide then the equations are

$$\text{iron} + \text{oxygen} \rightarrow \text{iron(III) oxide}$$
$$4Fe(s) + 3O_2(g) \rightarrow 2Fe_2O_3(s)$$

In terms of electron transfer, the iron has been oxidised because its atoms have lost electrons to form iron 3+ ions (see section 2.2).

$$Fe(s) \rightarrow Fe^{3+}(aq) + 3e^-$$

Prevention of rusting

Iron will not rust if it is kept in dry conditions or if it is kept in an atmosphere that does not contain oxygen. Oxygen and water are both needed for the iron to rust.

Rusting can be prevented by any process which makes a barrier between the iron or steel surface and the water and oxygen in the air.

Did you know?

A large steel structure like a bridge has to be continually repainted to maintain the protection and as soon as one end is reached, it is time to re-start at the other end.

Alloying

Special alloys of steel can be made by adding chromium or vanadium to the iron to reduce its rusting rate. These are called stainless steels and are used to make tools, kitchen utensils and work surfaces. Stainless steels are more expensive than ordinary steels.

Sacrificial protection

This is where a more reactive metal such as magnesium or zinc is attached to the iron or steel object. The more reactive metal will corrode in preference to the iron and steel and thus protect it. The more reactive metal will 'sacrifice' itself for the benefit of the iron or steel. **Sacrificial protection** method is used to protect large steel structures such as pylons and large ships.

Galvanising is a form of sacrificial protection. In this process iron or steel is covered with a layer of zinc. If the zinc coating gets damaged and the iron is exposed, it is the zinc which corrodes first.

Using the reactivity series to explain the prevention of rusting by sacrificial protection using magnesium or zinc

A fairly reactive metal such as magnesium or zinc can be used to prevent iron from rusting as quickly as it would normally. The iron or steel object has the more reactive metal attached to it either directly or connected by a conducting wire.

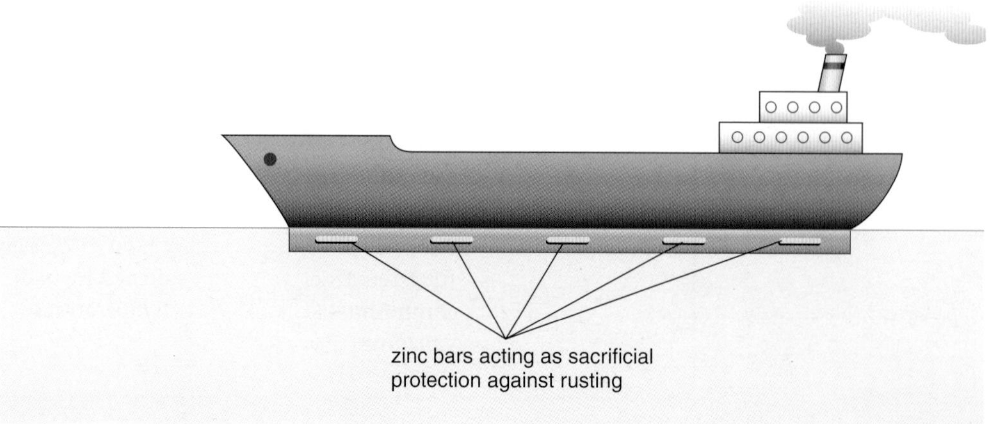

Figure 4.8 *Sacrificial protection of a ship's hull*

The more reactive metal will corrode in preference to the iron – its metal atoms will turn into metal ions.

$$Zn(s) \rightarrow Zn^{2+}(aq) + 2e^-$$

The released electrons will spread onto the iron and help to prevent a similar ionisation process occurring, which would be the first step in the rusting process. The zinc gradually corrodes away and 'sacrifices' itself to protect the iron.

Aluminium

Did you know?

Americans call aluminium 'ALUMINUM'. This was the original name suggested for the metal by Sir Humphry Davy.

Aluminium is a reactive metal – it is higher in the reactivity series than carbon. It cannot therefore be produced by smelting so electrolysis has to be used. Electrolysis is a very expensive process because it uses a lot of electricity. It is only possible to extract aluminium economically where there is a cheap source of electricity. The one cheap source of electricity is hydroelectric power.

The ore of aluminium is called **bauxite**. It is aluminium oxide (Al_2O_3).

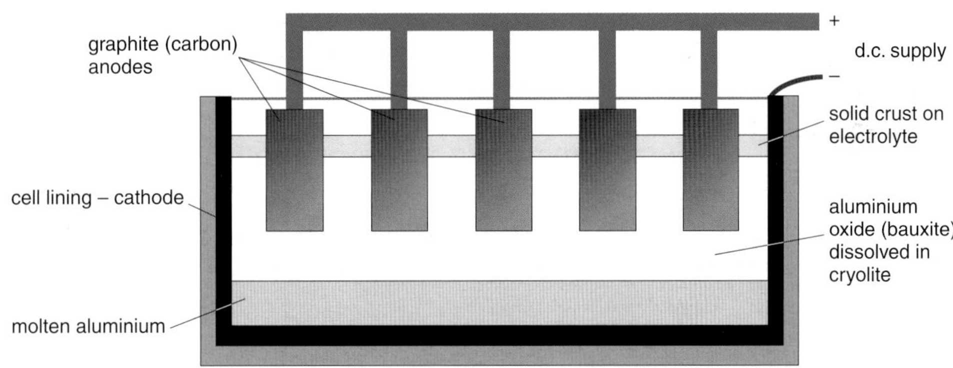

Figure 4.9 *Diagram of an aluminium electrolysis cell*

Aluminium oxide has a very high melting point (2050°C). To make the process safe to operate, another aluminium ore, **cryolite**, is used. Cryolite melts at a much lower temperature. Bauxite is dissolved in the molten cryolite. The cryolite does not get used up in the electrolysis process.

When an electric current is passed through the molten electrolyte, positively-charged aluminium ions (Al^{3+}) are attracted to the **cathode** (negative electrode). At the cathode, they gain electrons to form aluminium metal.

This is an example of reduction.

$$Al^{3+} + 3e^- \rightarrow Al(l)$$

The molten aluminium settles to the bottom of the tank and is syphoned off from the cell. At the **anode** (positive electrode), oxygen is produced.

The negatively charged ions lose electrons (oxidation).

$$O^{2-} - 2e^- \rightarrow O(g)$$

The carbon anode has to be replaced at regular intervals as it reacts with the oxygen to produce carbon dioxide.

In the electrolytic extraction of aluminium, oxidation and reduction take place. Together these are an example of a redox reaction.

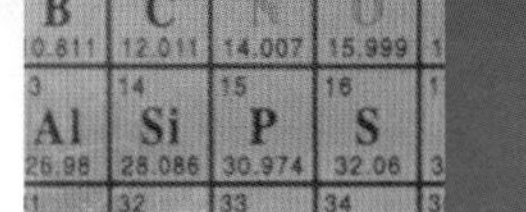

Uses of aluminium

Aluminium has a low density (2.7 g/cm^3) compared to iron (7.9 g/cm^3). It can be made very strong by alloying with other elements. This makes it ideal for use in the construction of aircraft. Aluminium is also a good conductor of heat so it can be used to make cooking utensils. Because of its good electrical conductivity and low density, it is used to make the high tension cables for the overhead power cables.

Figure 4.10 *Electrolytic cells for the production of aluminium*

Using the reactivity series to explain the unexpected behaviour of aluminium due to the formation of an impermeable oxide coating

The high position of aluminium in the reactivity series shows that it should be quite a reactive metal and would be expected to react with water and oxygen in the atmosphere. The normal uses to which it is put, for example window frames and lightweight bodies for railway engines and planes, show that this is not true.

The surface of the aluminium rapidly reacts with the oxygen in air to form a thin oxide layer, which protects the aluminium below it. This layer is impermeable and prevents any further reaction from taking place with more oxygen or water. The outer oxide layer acts as a protective layer. This layer differs from a rust layer on iron because it stays firmly in place and does not flake off exposing fresh metal surface.

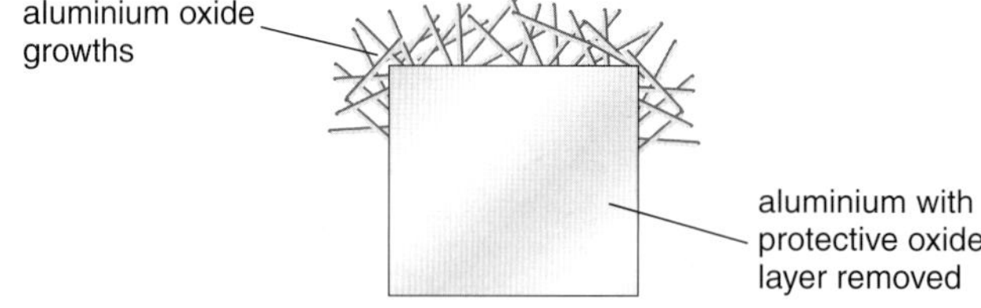

Figure 4.11 *The reaction of 'unprotected' aluminium with oxygen in air*

If the aluminium oxide layer is removed chemically with concentrated hydrochloric acid then a rapid reaction takes place between the newly exposed aluminium surface and oxygen in the air. Delicate, white, fern-like structures of aluminium oxide can be seen growing outwards from the surface of the aluminium. Aluminium is showing its real reactivity.

Using electrolysis to get pure copper

Copper is a good conductor of electricity. Electrolysis is used to make very pure copper.

In this process, copper atoms in the anode lose electrons and become copper ions. These ions pass into the copper sulphate solution. The anode gradually 'dissolves'.

At the cathode, copper ions from the solution gain electrons to become copper atoms. These atoms form on the surface of the cathode as very pure copper.

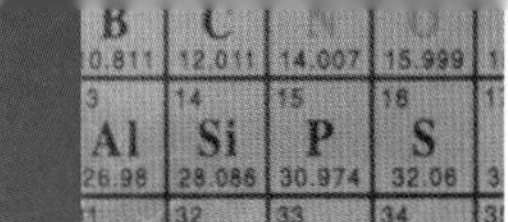

The electrolyte, a solution of copper(II) sulphate contains the following ions: Cu^{2+} and SO_4^{2-}, from the copper sulphate and H^+ and OH^-, from the water.

The copper **anode** has a positive charge because it has lost electrons to the cathode.

The OH^- and SO_4^{2-} ions are negatively charged so have had electrons added (see section 4.1).

Copper atoms in the anode form copper ions which move into the solution. As they do so they lose electrons. The hydroxide ions and sulphate ions remain in the solution. At the anode no substance is produced but copper is lost.

The reaction at the anode can be summarised as:
copper atoms – electrons → copper ions

$$Cu(s) - 2e^- \rightarrow Cu^{2+}(aq)$$

This is an example of oxidation.

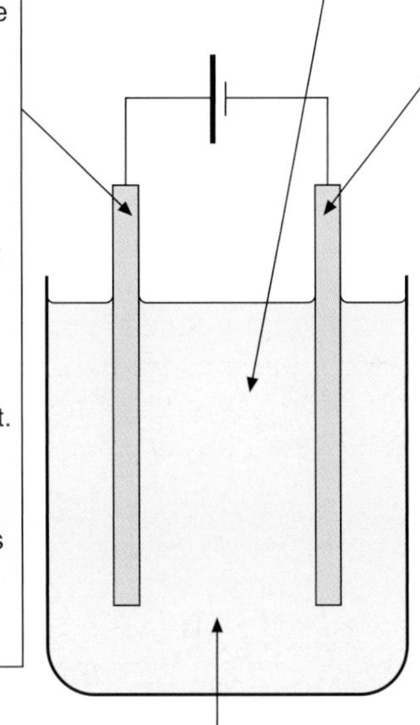

The copper **cathode** has a negative charge because it has gained extra electrons from the anode.

The H^+ and Cu^{2+} ions are positively charged so have had electrons removed (see section 4.1).

These postive ions are attracted to the negative electrode, the cathode, but because copper is less reactive than hydrogen (see section 4.2) the hydrogen ions remain as ions in the solution and only the copper ions form atoms. The cathode becomes coated with a layer of pure copper.

The reaction at the cathode can be summarised as:
copper ions + electrons → copper atoms

$$Cu^{2+}(aq) + 2e^- \rightarrow Cu(s)$$

This is an example of reduction.

The Cu^{2+} ions, the OH^- ions, the SO_4^{2-} ions and H^+ ions remain in solution.

Figure 4.12
Purifying copper by electrolysis

Did you know?

The industrial process of electroplating uses a similar process to the one used to purify copper. To get the metal being deposited to form on the article being plated, the conditions have to be very carefully controlled. Electroplating is used to put a thin layer of gold or silver on cheaper metals to make inexpensive jewellery. It is also used to cover metals like iron with other metals like chromium to prevent corrosion.

Completing and balancing half-equations for the reactions occurring at the electrodes during electrolysis

Half-equations provide information about the charge of the ion and the atomic or molecular nature of the product.

The table gives some information about the product at each electrode.

At the negative electrode (cathode)	At the positive electrode (anode)
Positively charged ions (for example, Cu^{2+}, H^+ and Al^{3+}) gain electrons	Negatively charged ions (for example, Cl^- and O^{2-}) lose electrons
The number of electrons gained equals the size of the charge on the ion	The number of electrons lost equals the size of the charge on the ion
Metallic elements are released as atoms (Cu and Al) Hydrogen released as molecules (H_2)	Gases released as molecules (Cl_2, O_2)

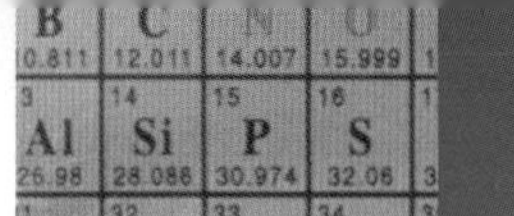

Example 1 Complete and balance the half-equation for the formation of chlorine at an anode.

$$Cl^- \quad - \quad e^- \longrightarrow Cl_2$$

Number of molecules of chlorine (Cl_2) produced = 1
So, number of chloride ions (Cl^-) involved = 2
So, number of electrons to be lost = 2
So, the completed and balanced equation will be

$$2Cl^- \quad - \quad 2e^- \longrightarrow Cl_2$$

Example 2 Complete and balance the half-equation for the formation of copper at a cathode.

$$Cu^{2+} \quad + \quad e^- \longrightarrow$$

Electrons will be gained.
1 atom of Cu will be produced.
So, number of copper ions (Cu^{2+}) involved = 1
So, number of electrons to be gained = 2
So, the completed and balanced equation will be

$$Cu^{2+} \quad + \quad 2e^- \longrightarrow Cu$$

Example 3 Complete and balance the half-equation for the formation of aluminium at a cathode.

$$Al^{3+} \quad + \quad e^- \longrightarrow$$

1 atom of Al will be produced
So, number of aluminium ions (Al^{3+}) involved = 1
So, number of electrons to be gained = 3
So, the completed and balanced equation will be

$$Al^{3+} \quad + \quad 3e^- \longrightarrow Al$$

Example 4 Complete and balance the half-equation for the formation of oxygen at an anode.

$$O^{2-} \quad - \quad e^- \longrightarrow$$

1 molecule of O_2 will be produced
So, number of oxide ions (O^{2-}) involved = 2
So, number of electrons to be lost = 4
So, the completed and balanced equation will be

$$2O^{2-} \quad - \quad 4e^- \longrightarrow O_2$$

Example 5 Complete and balance the half-equation for the formation of hydrogen at a cathode.

$$H^+ \quad + \quad e^- \longrightarrow H_2$$

1 molecule of H_2 is produced
So, number of hydrogen ions (H^+) involved = 2
So, number of electrons to be gained = 2
So, the completed and balanced equation will be

$$2H^+ \quad + \quad 2e^- \longrightarrow H_2$$

Did you know?

The demand for aluminium today is about 1000 times higher than it was 100 years ago but the cost of aluminium is about 20% of what it was 100 years ago.

Demand for raw materials

The demand for most raw materials is increasing. As the demand goes up, social, economic and environmental problems are created. For example:

- open-cast mining of ore uses up farmland and destroys habitats
- transportation of materials creates noise and air pollution
- mining ore and production processes produce dust and noise
- waste materials produce unsightly tips
- disused underground mines can collapse causing subsidence
- disused open-cast mines look unpleasant
- all disused sites can be dangerous.

Industrial processes

In this book several industrial processes are mentioned. These processes have been a benefit to people in a number of ways.

- New materials have been developed. A good example of this is the polymer (plastics) industry. This industry has grown as a result of advancements in the oil industry.
- Many industrial products have become easier to get and less expensive. This is because most of these processes take place in very large factories. Large factories can usually make products more efficiently than small factories and this keeps the cost of the product low. If products are made at low cost there will probably be more demand for them. This can mean that more factories will be built to meet that demand.
- The factories have provided employment with good wages.

But there are also pollution problems caused by these industrial processes.

- Big factories need a lot of land. This reduces the land available for growing food. It also affects the habitats of wildlife.
- Transportation – factories need to get their raw materials; they also need to send out the products they make. Both require transportation. Lorries produce noise and air pollution. Heavy lorries can also cause expensive damage to roads and buildings beside the roads. There are similar problems with rail transport, though they are not usually as bad. Ships may be used for transport – they can also pollute. Accidents at sea can be a serious problem because of the huge size of the ships and the amount of material they carry.
- Pollution – many of the processes produce dust and acidic or toxic fumes.
- Waste materials – most of the industrial processes produce substances that are not useful. These substances are often dumped in huge piles (tips), thus using up more land.

People's attitudes to scientific development is an important factor. The Haber process has been used for many years to produce cheap nitrogenous fertiliser (see section 3.2). Once this was considered an important scientific development – it allowed farmers to grow crops every year and increased food production in poor countries. Today, however, many people are unhappy about the use of artificial fertilisers. Because they were inexpensive, these fertilisers were used too much. Some of the fertilisers got into rivers and streams causing pollution (see section 3.2). Increasingly people are going back to natural fertilisers – they are growing and eating more 'organic' food.

Did you know?

In the last 50 or 60 years many scientific discoveries have been made. In many cases scientists believed these discoveries would be of great benefit to people. Only later was it found that there were problems that no-one has foreseen. Some of these discoveries, like the insecticide DDT and the CFCs that were used as aerosol propellants, harmed the environment. It is now much harder for scientists to convince people that new discoveries will be a benefit. The recent problem with genetically modified (GM) foods is an example of this.

Summary

- Rocks are usually a mixture of **minerals**.
- Minerals are usually chemical compounds.
- **Ores** are minerals that can be used as a source of useful materials.
- The **reactivity series** is a table of metals with the most reactive metals at the top and the least reactive metals at the bottom.
- The reactivity of the metal can be found by reacting it with water, acids and oxygen.
- A metal higher in the reactivity series will always remove oxygen from the oxide of a metal that is lower than it in the series.
- The **Thermit process** is a method of joining two lengths of railway track together using the greater reactivity of aluminium compared to iron.

- A metal higher in the reactivity series will displace a metal which is lower in the series from a solution of its metal sulphate.
- All **corrosion** processes involve a reaction between a metal and substances in the atmosphere.
- In **sacrifical protection** a more reactive metal such as magnesium or zinc can be used to prevent iron from rusting as quickly as it would normally.
- The surface aluminium rapidly reacts with the oxygen in air to form a thin oxide layer which protects the aluminium below it.
- Rusting is the corrosion of iron and happens in the presence of air and water.
- Rusting is the oxidation if iron because the iron has gained oxygen and its atoms have lost electrons to form iron 3+ ions.
- Rusting can be prevented by any process which makes a barrier between the iron or steel surface and the water and oxygen in the air.
- Metals less reactive than carbon can be extracted from their ore by **smelting**.
- Metals more reactive than carbon myst be extracted from their ore by **electrolysis**.
- Smelting is usually done by heating the ore with carbon.
- Smelting is usually done in a **blast furnace**.
- In the smelting of iron, iron ore (usually **haematite**), **coke** (fairly pure carbon made from coal) and limestone (calcium carbonate) are added at the top. Hot air is blown in at the bottom.
- Extracting a metal by electrolysis means passing an electric current through the molten ore.
- Aluminium is extracted by electrolysis.
- Half-equations can be used to describe electrolysis.
- Aluminium ore, **bauxite**, is dissolved in molten cryolite to lower its melting point.
- Pure metals can often be made stronger by adding other elements to make **alloys**.
- Industrial processes can cause social, economic and environmental problems.

Topic questions

1 For each of the following, say whether they are *best* described as mineral, ore or rock.
  a) granite
  b) bauxite
  c) haematite

2 Would you use smelting or electrolysis to get the following metals from their ores?
  a) zinc
  b) calcium
  c) iron
  d) potassium

3 a) What three substances are put in the top of a blast furnace?
  b) Which one of these is the ore?
  c) Which one of these is the reducing agent?
  d) What is the purpose of the third substance?
  e) What other substance has to be put into the blast furnace?

4 The gases carbon monoxide and carbon dioxide are made in the blast furnace.
  a) Write a chemical equation for a reaction in a blast furnace that produces carbon dioxide.
  b) Write an equation for a reaction in a blast furnace that produces carbon monoxide.
  c) Which of these gases will reduce iron ore to produce iron?

5 Steel is an alloy of iron and carbon.
  a) What is meant by the term ‘alloy’?
  b) What effect does 1% of carbon have on iron?
  c) What effect does 4% of carbon have on iron?
  d) What is the alloy of iron which contains 4% carbon called?

6 a) Bauxite is the main ore of which metal?
  b) What other ore of this metal is mixed with bauxite?
  c) What are the advantages of using a mixture of ores?

7 The cables that are used to carry electricity along overhead power lines are made of pure aluminium. The aluminium is not strong enough by itself and is wrapped round a central core of steel. What properties of aluminium make it a good choice for high tension cables?

8 Complete and balance the following half-equations for products produced at the electrodes during electrolysis.
  a) For chlorine produced at a positive electrode
  $2Cl^- \quad - \quad e^- \longrightarrow$
  b) For copper produced at a negative electrode
  $Cu^{2+} \quad + \quad e^- \longrightarrow \quad Cu$
  c) For aluminium produced at a cathode
  $Al^{3+} \quad e^- \quad \longrightarrow$
  d) For oxygen produced at an anode
  $O^{2-} \quad e^- \quad \longrightarrow \quad O_2$
  e) For hydrogen produced at a negative electrode
  $H^+ \quad e^- \quad \longrightarrow$

4.3 Useful products from rocks

Co-ordinated	Modular
11.5	06 (10.6)

Limestone and its uses

Limestone landscapes are considered by many people to provide the most spectacular landscapes in England. But because limestone is such a useful raw material there are difficult environmental decisions that need to be made when quarrying for limestone.

Figure 4.13
A limestone pavement in Malham, Yorkshire

This diagram shows some uses of limestone.

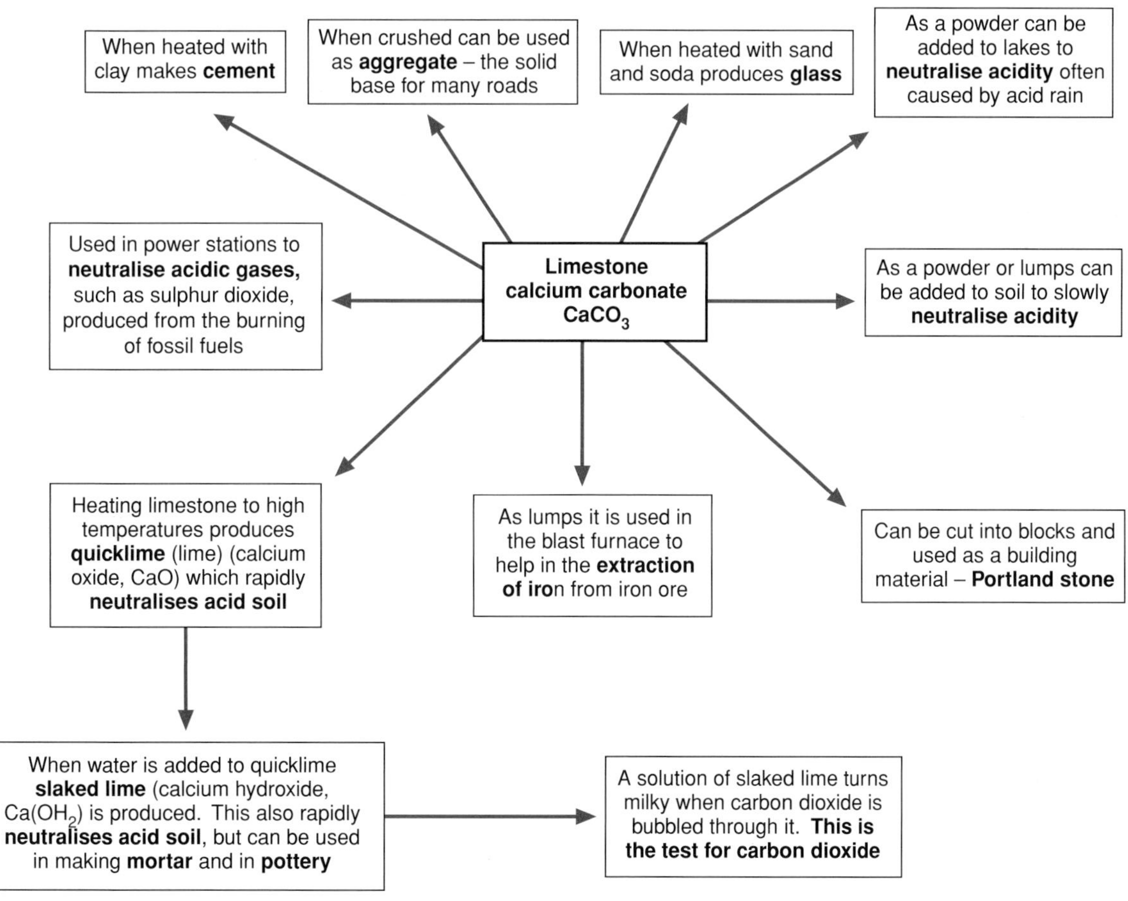

Figure 4.14
Some uses of limestone

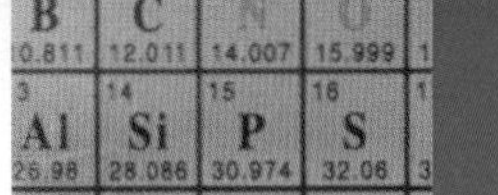

Heating limestone

For many centuries limestone has been heated in lime kilns to provide quicklime and slaked lime which farmers use to reduce soil acidity. The temperature inside these kilns is about 1200°C.

When limestone – calcium carbonate – is heated strongly it breaks down into calcium oxide and carbon dioxide:

calcium carbonate	→	calcium oxide	+	carbon dioxide
$CaCO_3$ (s)	→	CaO (s)	+	CO_2 (g)
		(quicklime/ lime)		

This reaction is an example of **thermal decomposition** – the breaking down of a substance by heating (see section 2.3).

Did you know?

If lime is heated strongly it will produce an intense white light. This effect was used in theatres, hence the phrase 'in the limelight'.

Figure 4.15
Limelight being used in theatre spotlights

If water is added to calcium oxide there is a vigorous reaction during which much heat is released and calcium hydroxide is produced.

calcium oxide	+	water	→	calcium hydroxide
CaO (s)	+	H_2O (l)	→	$Ca(OH)_2$ (s)
				(slaked lime)

Large quantities of calcium hydroxide are used in industry as an alkali.

The laboratory test for carbon dioxide

A solution of calcium hydroxide in water is called **limewater**. This is used as a test for carbon dioxide. As carbon dioxide is bubbled through limewater the solution goes milky due to the formation of insoluble calcium carbonate.

calcium hydroxide	+	carbon dioxide	→	calcium carbonate	+	water
$Ca(OH)_2$(aq)	+	CO_2 (g)	→	$CaCO_3$(s)	+	H_2O(l)

Limestone and cement

Cement is made by heating a mixture of crushed limestone, clay and shale at a temperature of about 1600°C. The heating is carried out in a rotary kiln more than 150 m long and 4m in diameter. The resulting clinker is then cooled and ground down into a very fine powder – cement.

Did you know?

The name 'Portland cement' was first used in 1824 by a British cement maker called Joseph Aspdin. He realised that the concrete made from Portland cement looked like the building material Portland stone.

Portland cement mixed with water and sand or gravel forms concrete. The concrete hardens as the water and cement powder react chemically.

Limestone and glass

Glass was thought to have been first made nearly 4000 years ago. Most of the glass used to make bottles, lamp bulbs, window and plate glass is called soda-lime glass. This glass is made by heating a mixture of lime, soda (sodium carbonate, Na_2CO_3) and sand (silica).

Summary

- Limestone (calcium carbonate) is quarried and used as a building material.
- Powdered limestone is used to neutralise acidity in lakes and soils.
- Limestone heated in kilns breaks down into quicklime (calcium oxide) and carbon dioxide. This is an example of **thermal decomposition.**
- Quicklime reacts with water to form slaked lime (calcium hydroxide) which is used to neutralise soil acidity.
- Roasting powdered limestone with powdered clay in a rotary kiln produces cement.
- When cement is mixed with water, sand and crushed rock, a slow chemical reaction produces concrete.
- Glass is made by heating a mixture of limestone, sand and soda (sodium carbonate).
- Carbon dioxide is a gas. It will turn limewater milky.

Topic questions

1 Why is limestone such a useful raw material?

2 Write down the chemical names and formulae for each of the following:
   a) limestone
   b) quicklime
   c) slaked lime
   d) soda

3 How is limestone used in power stations?

4 Explain why the heating of limestone is an example of thermal decomposition.

5 Describe the limewater test for carbon dioxide. Use a word equation in your answer.

6 How is cement made?

7 What is concrete?

8 What are the three raw materials needed to make window glass?

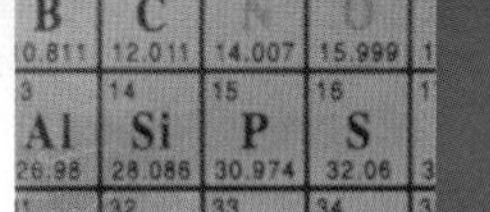

4.4

Co-ordinated	Modular
11.3	06 (10.3)

Useful products from crude oil

Crude oil

Crude oil and natural gas are **fossil fuels** that were formed from small animals and plants which died and were buried under sediment well over 100 000 000 years ago. The organic matter decomposed (broke down) to form crude oil and natural gas which remained trapped in the rocks.

Crude oil and natural gas are examples of **non-renewable** (or **finite**) **resources**. Coal is a further example. Once non-renewable resources are used up, they cannot be replaced.

Crude oil is a mixture of substances. Most of these substances are **hydrocarbons** (compounds which contain hydrogen and carbon only).

A mixture is made if two or more elements or compounds come together but do not chemically join. In a mixture the chemical properties of each substance remain unchanged. Because their properties are unchanged the substances in a mixture can be separated by physical methods such as distillation.

Did you know?

Most of the oil and gas produced by fossilisation has seeped up through the rock and escaped into the atmosphere. It is only when it has been trapped under a layer of non-porous rock that it can be extracted.

Fractional distillation

Fractional distillation is used to separate two or more liquids that are mixed together, for example ethanol and water. Fractional distillation is the equivalent of a series of simple distillations and is very efficient. A fractionating column is put in between the boiling liquid mixture and the condenser.

The liquid mixture is heated until it boils and the vapour formed rises up into the fractionating column. The vapour condenses and re-boils many times as it moves up the column. Each time this happens, the composition of the vapour changes to contain more of the most volatile liquid (the one with the lowest boiling point). If the column is long enough, vapour from each of the liquids in the mixture gradually separates and makes its way successively up the column. The temperature of each of the separated vapours is shown by the thermometer before it condenses. As each liquid condenses in turn, it can be collected in separate flasks. The liquid with the lowest boiling point will come out first.

Figure 4.16 *Fractional distillation of a liquid mixture*

Separating crude oil

Crude oil can be separated into its various useful components by fractional distillation.

The different fractions from crude oil have different properties and therefore different uses.

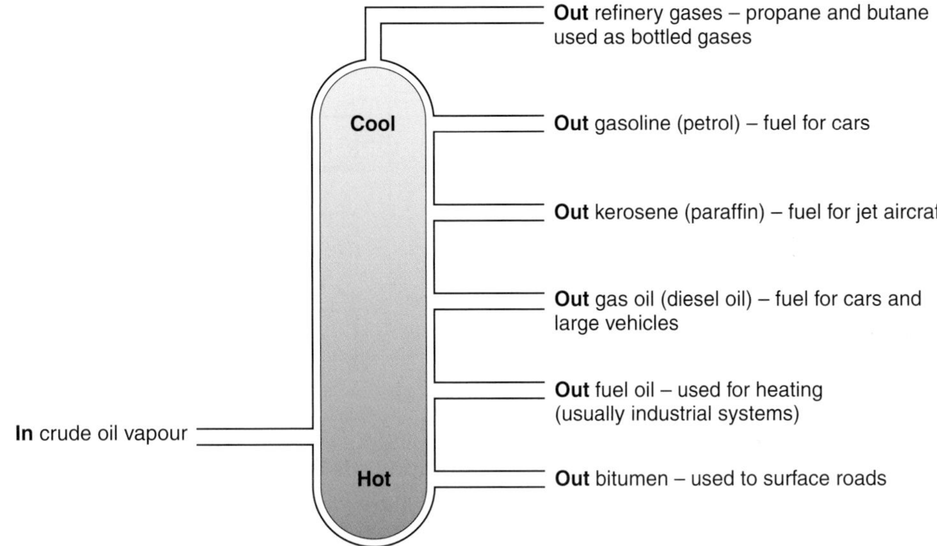

Figure 4.17 *Simplified diagram of the fractional distillation of crude oil*

Products that condense lower down the fractional distillation column have a higher boiling point. The higher the boiling point, the more carbon atoms there are in each molecule.

Figure 4.18 *Table comparing the properties of some of the fractions of crude oil*

Fraction	Approx. no. of carbon atoms	Approx. boiling point/°c	Appearamce	What happens
gasoline (petrol)	4–10	less than 100	almost colourless liquid with low **viscosity***	highly flammable; has an almost colourless, smokeless flame
gas oil (diesel)	16–20	about 200	yellow liquid with a higher viscosity	flammable; burns with a yellow, smoky flame
bitumen	50+	over 350	black substance – so viscous it's almost a solid	can only be made to burn if heated to a high temperature and sprayed as fine droplets

***Viscosity** is a measure of how runny a liquid is. Water has a low viscosity and treacle has a high viscosity.

Huge quantities of oil are used. Most of it by the industrialised nations of North America and Europe. Much of the oil is transported to these countries in large ocean-going tankers. Oil leaks from these ships – this is a problem if a tanker is involved in an accident. There are many examples where this has happened. Oil does not mix with water, it floats on top producing an 'oil slick'. Oil slicks are a serious environmental hazard. They kill wildlife at sea and can pollute beaches damaging the habitats of many creatures.

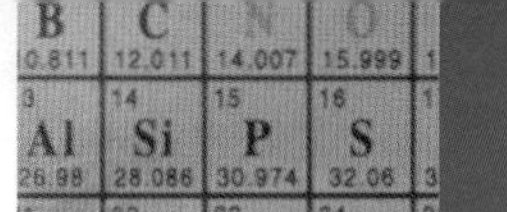

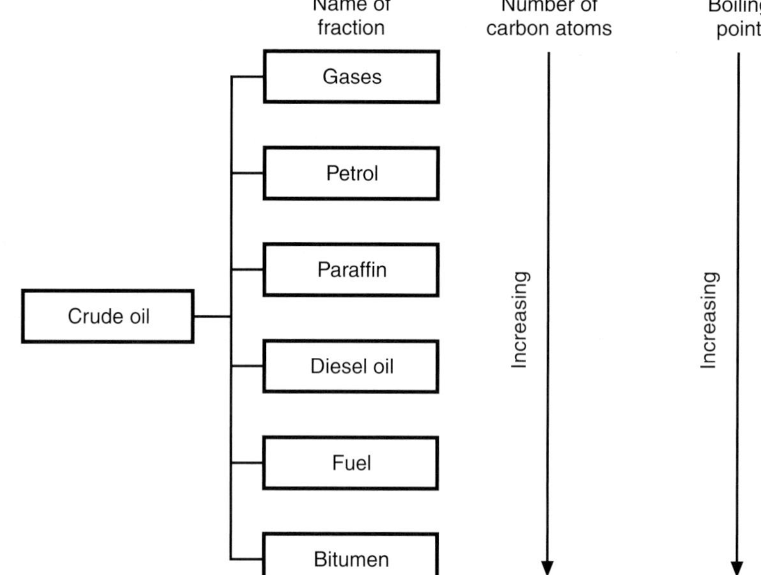

Figure 4.19 *The main fractions from crude oil*

Did you know?

About 2 tonnes of North Sea crude oil are used up each year for every person in the UK.

Hydrocarbons

Hydrocarbons are compounds containing only HYDROgen and CARBON. Natural gas and most of the substances in crude oil are hydrocarbons. All hydrocarbons are **flammable**. This means that the reaction of these compounds with oxygen is exothermic (see section 6.1). This is why hydrocarbons can be used as fuels. The smaller the size of the molecule, the easier it burns (see Figure 4.18 above). If there is enough air, the products of burning are carbon dioxide and water (vapour).

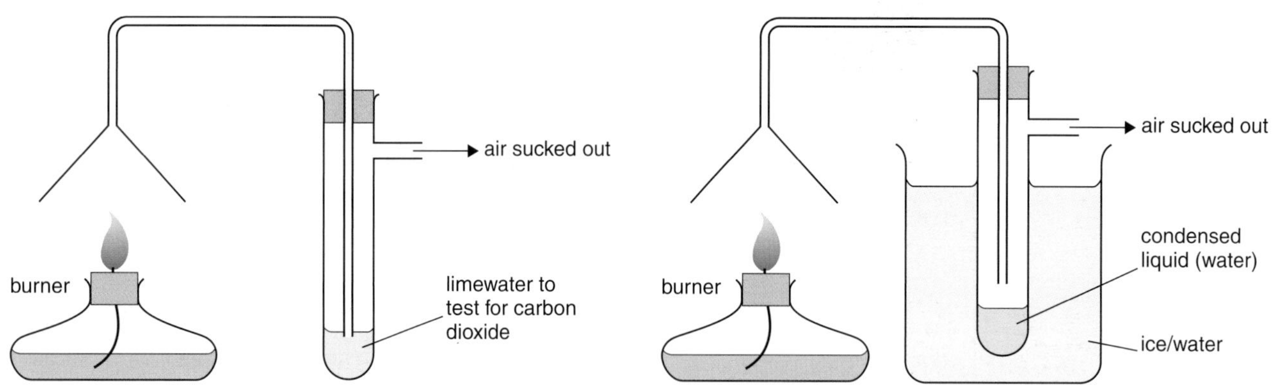

Figure 4.20 *Apparatus to show that burning hydrocarbons produces carbon dioxide and water vapour*

The apparatus above can be used to show that burning a hydrocarbon produces carbon dioxide and water. Figure 4.21 gives the tests for carbon dioxide and water (see section 6.2).

Substance	Test for substance	Positive result
carbon dioxide	mix with (or bubble through) limewater	**limewater** goes 'milky'
water	1 add to blue cobalt chloride* 2 add to white (anhydrous) copper(II) sulphate* 3 check the boiling point of the liquid*	cobalt chloride turns pink copper(II) sulphate turns blue water boils at 100°C

*Tests 1 and 2 will show that the liquid contains water. Test 3 will show that the liquid is pure water. (Impurities raise the boiling point above 100°C).

Figure 4.21 *Tests for carbon dioxide and water*

Unfortunately most hydrocarbon fuels not only contain carbon and hydrogen but often also contain sulphur. When these fuels are burnt, carbon dioxide, water and sulphur dioxide are produced.

If there is not enough air, the **combustion** of the hydrocarbon may be incomplete. If this happens the products of combustion will include carbon monoxide and carbon. If carbon is produced, the flame will be yellow and smoky.

The word equations for burning methane (the main gas present in natural gas) when there is not enough air are:

methane + (not enough) oxygen → carbon + water vapour

and

methane + (not enough) oxygen → carbon monoxide + water vapour

If there is enough air the reaction is:

$$\text{methane} + \text{oxygen} \rightarrow \text{carbon dioxide} + \text{water vapour}$$
$$CH_4(g) + 2O_2(g) \rightarrow CO_2(g) + 2H_2O(g)$$

To completely burn one molecule of methane takes two molecules of oxygen (see section 2.4). To completely burn one molecule of the substances in petrol would take about 10 molecules of oxygen. Over 70 molecules of oxygen are needed to burn one of the molecules present in bitumen. This means that the bigger the molecule, the more likely it is that combustion will not be complete. This is why hydrocarbons with large molecules burn with smoky, yellow flames. The black smoke is carbon and the yellow colour of the flame is caused by the carbon particles glowing.

Did you know?

If natural gas is burned in a bunsen burner with the air hole closed, the flame is yellow because there is not enough air. With the air hole open, much more air can get in and the gas is burned completely. When the gas is burned completely, all the available energy of combustion is released. This is why the blue flame is hotter than the yellow flame.

How the burning of hydrocarbon fuels affects the environment

Burning hydrocarbon fuels causes a number of problems.

- The production of large amounts of carbon dioxide has caused the level of carbon dioxide in the atmosphere to rise. The result is an increase in **global warming** – the **greenhouse effect**.
- If hydrocarbons are burned in a poor supply of air, carbon monoxide is produced. Carbon monoxide is a colourless gas that has no smell. It is highly poisonous.
- Sulphur dioxide is produced because most hydrocarbon fuels contain sulphur. Sulphur dioxide dissolves in the water in the air to form acid rain.

Did you know?

Carbon monoxide kills because it combines with haemoglobin in the blood. This stops the haemoglobin binding to oxygen, so the body's cells get starved of oxygen and stop working. As little as 0.1% of carbon monoxide can be fatal.

Alkanes

Alkanes are hydrocarbons. The simplest of the alkanes is methane, formula CH_4.

All alkanes have the general formula (C_nH_{2n+2}). In the molecule, all the bonds are covalent (see section 1.2). The **structural formula** of methane is:

```
     H
     |
 H — C — H
     |
     H
```

Did you know?

The methane molecule isn't really 'flat', it is in the shape of a tetrahedron. A tetrahderon has four faces. All the faces are equilateral triangles. The diagrams below are of two different 'models' of a methane molecule.

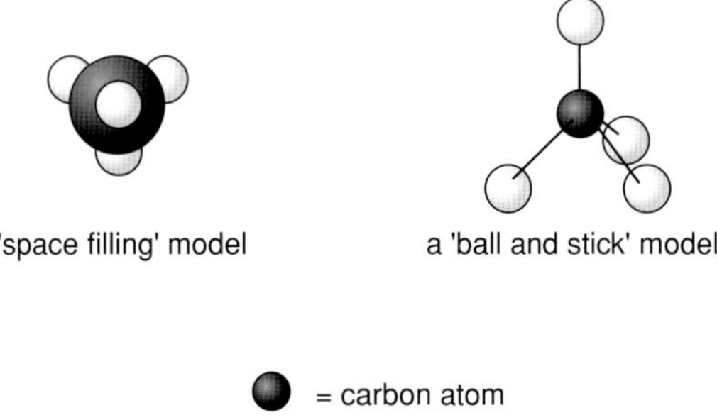

= carbon atom

= hydrogen atom

The diagrams above show the shape of a methane molecule.

Figure 4.22 gives the names and formulae of the first three alkanes of the alkane family.

Name of alkane	Molecular formula	Structural formula
methane	CH_4	H | H — C — H | H
ethane	C_2H_6	H H | | H — C — C — H | | H H
propane	C_3H_8	H H H | | | H — C — C — C — H | | | H H H

Figure 4.22 *The molecular and structural formulae of the first three alkanes*

Alkenes

Alkenes are also hydrocarbons. The simplest of these alkenes is ethene which has the formula C_2H_4. Poly(ethene), common name polythene, is made from ethene.

All alkenes have the general formula C_nH_{2n}. In the molecule all the bonds are covalent (see section 1.2).

Name of alkene	Molecular formula	Structural formula
ethene	C_2H_4	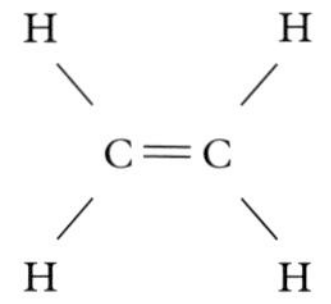 H₂C=CH₂ (each C bonded to two H)
propene	C_3H_6	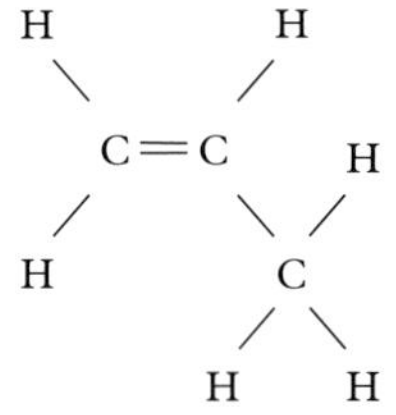 H₂C=CH–CH₃

Figure 4.23
The molecular and structural formulae of the first two alkenes

The test for alkenes

Alkenes have a carbon–carbon double covalent bond (see section 1.2). Unlike a carbon–carbon single bond, the double bond is not particularly strong. It can be attacked by some chemicals. This means that alkenes are more reactive than alkanes. For example, **bromine water** will react with ethene but not with ethane. The bromine water loses its colour in the reaction.

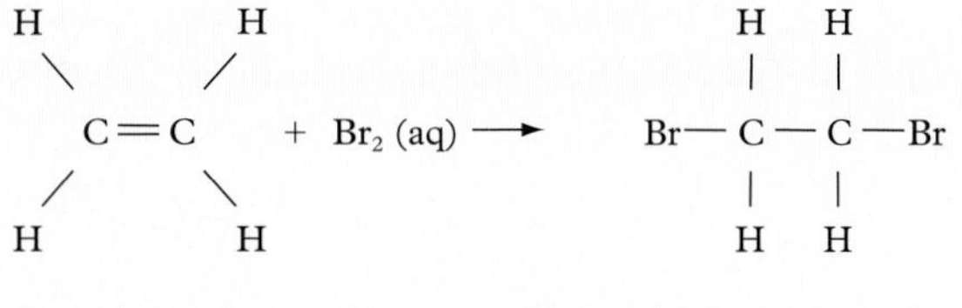

$$H_2C{=}CH_2 + Br_2\,(aq) \longrightarrow Br{-}CH_2{-}CH_2{-}Br$$

ethene + bromine water (yellow) → dibromoethane (colourless)

Alkenes decolourise bromine water, alkanes do not. This reaction is used as a test to tell the difference between alkanes and alkenes.

Compounds like alkenes which have a C=C double bond are called **unsaturated hydrocarbons**. Alkanes have no C=C double bond, only C—C single bonds. They are called **saturated hydrocarbons**.

Cracking

If the molecules in crude oil are heated strongly, they can break down into smaller molecules, some of which are useful as fuels. Industrially a catalyst is used to speed this process up. This is an example of a thermal decomposition reaction. The thermal decomposition of molecules in crude oil is called **cracking**. It is possible to carry out catalytic cracking on a small scale in the laboratory.

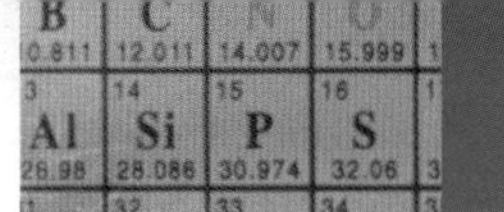

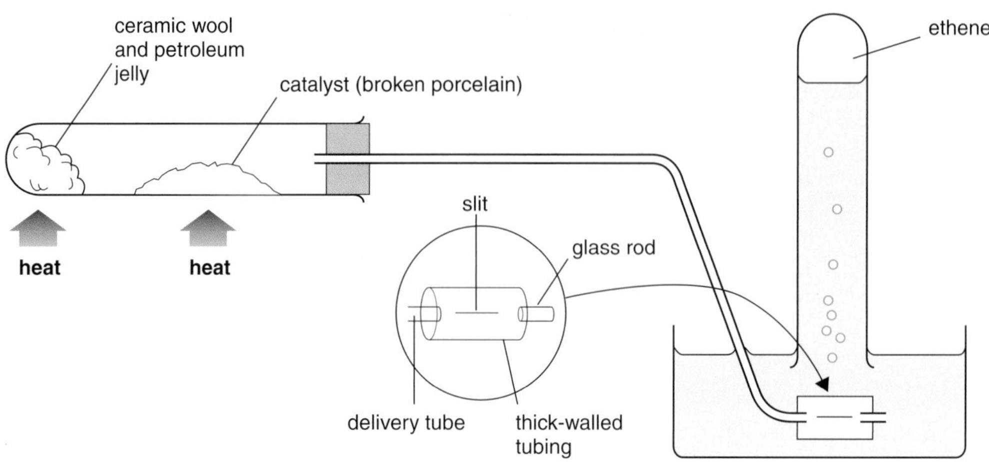

Figure 4.24
Apparatus for producing ethene in the laboratory by cracking petroleum jelly

When alkanes are cracked, some of the products are alkenes. In the example below the alkene produced is ethene.

$$\text{alkane} \rightarrow \text{alkane} + \text{alkene}$$
$$C_{10}H_{22} \rightarrow C_8H_{18} + C_2H_4$$
$$\text{decane} \rightarrow \text{octane} + \text{ethene}$$

Cracking is useful because it allows the large molecules of some unwanted products in crude oil to be broken down into more useful substances which have smaller molecules. Some of these products can be used as fuels.

Polymerisation

Polymers are large molecules made by joining smaller molecules together. The process is called **polymerisation**. The smaller molecules are called **monomers**. Plastics are polymers. One of the commonest polymers is poly(ethene) or polythene. It is made by polymerising molecules of ethene.

Polymerisation reactions of this type occur when unsaturated molecules (alkenes) join together to produce saturated alkanes with large molecules. Figure 4.25 shows the polymerisation of ethene to produce poly(ethene).

Figure 4.25
The polymerisation of ethene

n molecules of ethene → have their C=C bond broken → and link up to form a chain containing n C_2H_4 units

In this polymerisation reaction the alkene molecules join together. Polymerisation reactions of this type are called **addition polymerisation** reactions. This is because the molecules add on to each other.

Some of the products of cracking are used to make plastics. Poly(ethene) is used to make 'plastic' bags and bottles. Poly(propene) tends to be stronger than poly(ethene). It is used to make crates (e.g. milk crates) and ropes.

Did you know?

Polymers can be thermoplastic or thermosetting. Thermoplastic polymers soften when they are heated so they can be shaped easily. Most thermoplastics can be recycled. Thermosetting polymers are 'cured' (set) by heating. These polymers are not softened by heating. Thermosetting polymers cannot usually be recycled easily.

How the disposal of plastics affects the environment

There are three main ways by which most waste is dealt with. It can be dumped in landfill sites, burnt (incinerated) or recycled.

The demand for polymer materials (plastics) is increasing. Because the plastics are fairly cheap they are often thrown away when finished with. Many landfill sites discourage the disposal of plastics because most plastics are not **biodegradable**. Unlike wood, paper, wool and cotton, plastics do not break down by the action of micro-organisms, and unlike metal, plastics do not corrode over time. Recently some biodegradable plastics have been developed.

Figure 4.26
A landfill site

Figure 4.27
An incinerator

Figure 4.28
Collecting skip for plastic bottles

Incineration involves the burning of waste in a very hot furnace. Some people consider that the burning of many plastics, especially those made of polyvinyl chloride (PVC), produces a harmful group of chemicals called dioxins and heavy metals. The dioxins are thought to be associated with birth defects and some kinds of cancers. The advantage of the incineration of plastics is that it provides a relatively cheap source of energy for the community.

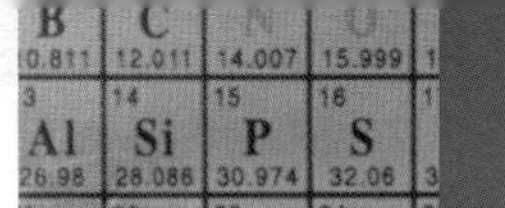

Recycling is now being encouraged in many parts of the world as it seems to be the most practical method to deal with the problem of disposal. For most plastics the process of recycling is fairly cheap and straightforward.

Summary

- **Crude oil** is another useful substance extracted from the Earth.
- Crude oil is the fossilised remains of small animals and plants.
- Crude oil is mainly made of **hydrocarbons**.
- Crude oil can be separated into its various hydrocarbons by **fractional distillation**.
- Hydrocarbons are compounds that contain hydrogen and carbon only.
- Hydrocarbons burn in plenty of air to produce water and carbon dioxide.
- Hydrocarbons burn in a limited air supply to produce water vapour, carbon monoxide and carbon.
- Carbon dioxide is a gas. It will turn limewater milky.
- Hydrocarbons can be **alkanes** or **alkenes**.
- Alkanes have C—C single bond.
- Alkanes are called **saturated hydrocarbons**.
- Alkenes have a C=C double bond.
- Alkenes are called **unsaturated hydrocarbons**.
- Large hydrocarbon molecules can be broken down into smaller ones by a process called **cracking**.
- Small alkene molecules can be joined together to make very large alkane molecules by the process of **polymerisation**.

Topic questions

1 The table opposite lists the properties of some of the hydrocarbons obtained from crude oil. (The hydrocarbons are not listed in any particular order).

Write the letters, A, B or C for each of the following. (You may use each letter once, more than once or not at all.)

Which hydrocarbon:

a) could be petrol?
b) could be bitumen?
c) has the highest boiling point?
d) has seven or eight carbon atoms in its molecule?
e) could be used to power a motor car?

Hydrocarbon	Appearance	Viscosity	What happens when the hydrocarbon is burned
A	black liquid	very high	very difficult to get it to burn at all
B	colourless gas	very low	burns very easily with a blue, smokeless flame
C	colourless liquid	medium	burns easily with smoky, yellow flame

2 Complete the following sentences.

When hydrocarbon fuels are burnt in plenty of air, they produce the gas carbon dioxide and ________. The test for carbon dioxide is to add the gas to ________. If carbon dioxide is present, the liquid goes ________ If there is not enough air present when the hydrocarbon is burnt, the gas ________ can be produced. When another fuel was burned in plenty of air, *only* carbon dioxide was produced. The fuel used was ________.

3 a) For each of the following hydrocarbons, state if the molecule is saturated or unsaturated.

i) methane
ii) ethene
iii) propane

b) Which of the hydrocarbons named in a) is the main substance in natural gas?

c) Which of the hydrocarbons named in a) would decolourise bromine water?

d) Which of the hydrocarbons named in a) would have the general formula C_nH_{2n}?

e) Write a balanced chemical equation for the complete combustion of methane.

f) Draw a structural formula for one alkane and one alkene mentioned in part a).

4 a) What is the name given to the process used to break large hydrocarbon molecules into smaller ones?

b) Why is it useful to be able to break large hydrocarbon molecules into smaller ones?

c) When the process is applied to an alkane are the products:
A all alkanes?
B all alkenes?
C a mixture of alkanes and alkenes?

5 Complete the following sentences using the words or phrases in the box.

large small poly(ethene) polygon clothes plastic plastic bottles rope

Polymers are ________ molecules made by joining ________ molecules together. One common polymer is ________ . This polymer is used to make ________ .

Examination questions

1 a) Complete the sentences.

i) A mineral, or a mixture of minerals, from which a metal can be extracted is called an ________. *(1 mark)*

ii) The reactivity series for some metals is:

potassium
sodium
calcium
magnesium
aluminium
zinc
iron
copper
gold

The name of the method used to extract potassium from the mineral potassium chloride is ________. *(1 mark)*

b) The diagram shows the type of furnace used to extract iron.

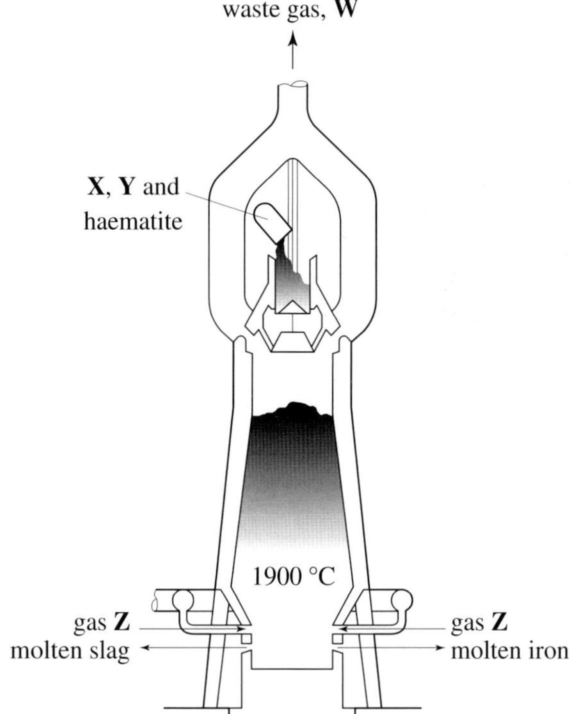

i) What is the name of this type of furnace? *(1 mark)*

ii) Name the main gas, other than nitrogen, in the waste gas, W. *(1 mark)*

iii) Name the raw materials shown as X, Y and Z. *(3 marks)*

iv) Carbon monoxide gas forms in the furnace. Balance the chemical equation for a reaction which produces it.

$$\text{—}\ C(s) + O_2(g) \rightarrow \text{—}\ CO(g)$$

(1 mark)

c) Carbon monoxide reacts with iron(III) oxide. This is the chemical equation for the reaction which occurs.

$$Fe_2O_3(s) + 3CO(g) \rightarrow 2Fe(l) + 3CO_2(g)$$

i) What does the symbol (l) mean? *(1 mark)*

ii) Complete the **two** spaces with the names of the chemicals.
In this reaction ________ is oxidised to ________. *(1 mark)*

2 Aluminium can be extracted from its ore bauxite. Bauxite contains aluminium oxide, Al_2O_3, which is purified and then processed as shown.

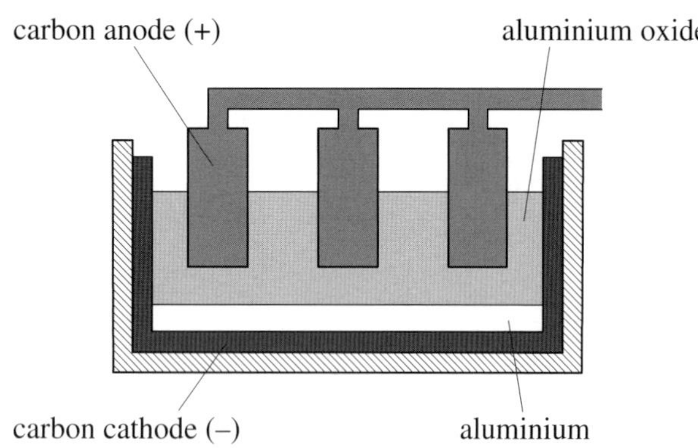

a) The word equation for the extraction of aluminium is shown.

aluminium oxide → aluminium + oxygen

i) Write the balanced chemical equation for this reaction. *(3 marks)*

ii) Describe how aluminium is extracted in this process. *(3 marks)*

b) During the extraction carbon dioxide gas is produced. Suggest why. *(2 marks)*

3 The high demand for petrol (octane) can be met by breaking down longer hydrocarbons, such as decane, by a process known as cracking.

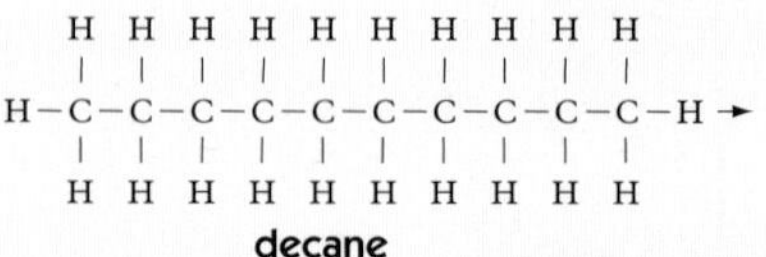

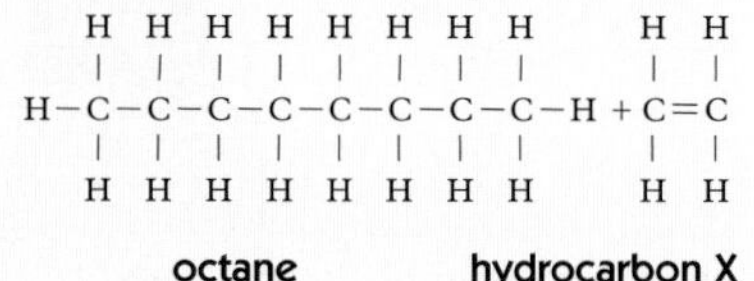

a) Apart from heat, what is used to make the rate of this reaction faster? *(1 mark)*

b) Octane is a **hydrocarbon**.

i) What does **hydrocarbon** mean? *(1 mark)*

ii) Give the molecular formula of octane. *(1 mark)*

c) The hydrocarbon **X** is used to make poly(ethene).

i) What is the name of **X**? *(1 mark)*

ii) What is the name of the process in which **X** is changed into poly(ethene)? *(1 mark)*

4 The table gives information about some alkanes

Name of alkane	Number of carbon atoms in each molecule	boiling point of alkane (°C)
methane	1	−161
propane	3	−42
butane	4	0
pentane	5	
hexane	6	69
octane	8	126
decane	10	

a) Draw a graph of boiling point against number of carbon atoms in the molecule for these alkanes. (Use graph paper.)
Draw your graph so that it can be extended to allow you to find the boiling point of the alkane with ten carbon atoms in each molecule. *(3 marks)*

b) Use your graph to find the boiling point of:

i) pentane

ii) decane. *(2 marks)*

5 The label has been taken from a tube of *Humbrol Polystyrene Cement*, a glue used in model making.

HUMBROL

Polystyrene Cement

HARMFUL

Paint product contains 1.1.1 TRICHLOROETHANE

• Keep container tightly closed. Harmful by inhalation, in contact with skin and if swallowed. Avoid contact with eyes. Keep out of reach of children.

• For use on all polystyrene plastic except expanded or foam. Specially recommended for plastic kits. Thinly coat each surface, press together. To remove cement from fabrics use Humbrol Universal Cleaner.

HUMBROL LTD., HULL, ENGLAND.

a) The solvent used is 1,1,1-trichloroethane. The structural formula of this molecule is:

```
    Cl   H
    |    |
Cl — C — C — H
    |    |
    Cl   H
```

i) What do the lines between the atom represent? *(1 mark)*

ii) State whether 1,1,1-trichloroethane is saturated or unsaturated. Give **one** reason for your answer. *(1 mark)*

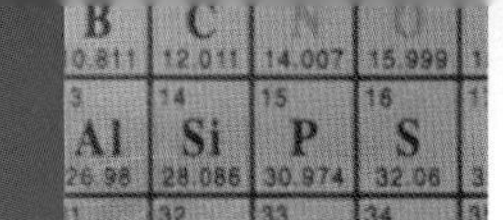

iii) 1,1,1-trichloroethane is being replaced in favour of a 'better' solvent. Use information on the label to help you to suggest why. *(1 mark)*

b) Polysytrene is a plastic. Plastics are polymers which are made by the process of polymerisation.

i) What is meant by polymerisation? *(2 mark)*

ii) The table gives information about monomers and the polymers made from them. Complete the table. *(3 marks)*

MONOMER		POLYMER	
name	formula	name	formula
ethene	$H_2C{=}CH_2$		$\left(\begin{array}{cc} H & H \\ | & | \\ -C & -C- \\ | & | \\ H & H \end{array}\right)_n$
styrene		polystyrene	$\left(\begin{array}{cc} H & H \\ | & | \\ -C & -C- \\ | & | \\ H & C_6H_5 \end{array}\right)_n$
chloroethene	$H_2C{=}CHCl$	poly(chloro-ethene)	

Chapter 5
Patterns of behaviour

Key terms acids • alkali • alkali metals • anode • atomic number • bases • cathode • displacement • electrolysis • electrolyte • galvanising • group • halide • halogens • hydrogen • hydrogen ion • hydroxide ions • indicators • inert • ions • neutral • neutralisation • noble gases • period • periodic table • pH • pH scale • transition elements • transition metal • universal indicator

5.1 The development of the periodic table

Co-ordinated	Modular
11.11	08 (10.13)

The story behind the periodic table of the elements

In order to understand why the **periodic table** came into existence it is important to be aware of some of the work on elements that was being carried out between 1770 and 1810.

The work of a French chemist Antoine Laurent Lavoisier (1743–1794) had helped scientists move away from Aristotle's idea of the four elements being air, earth, fire and water by developing the idea of a chemical element. He was the first scientist to define an element as being a substance that could not be broken down by chemical methods.

Lavoisier believed that elements could be divided into four groups according to their chemical behaviour. His table of elements was published in 1789.

Figure 5.1 *Table to show Lavoisier's attempt at classifying some simple substances. The modern names are given in brackets*

Acid-making elements	Gas-like elements	Metallic elements		Earthy elements
charcoal (carbon)	azote (nitrogen)	cobalt		argilla (aluminium oxide)
phosphorus	caloric (heat)	mercury		barytes (barium sulphate)
sulphur	hydrogen	copper	nickel	lime (calcium oxide)
	light	gold	silver	magnesia (magnesium oxide)
	oxygen	iron	tin	silex (silicon dioxide)
		lead	zinc	

Lavoisier's table includes some substances we know as compounds. He classified them as elements because the chemical method needed to break them down was not known at that time.

In 1808 John Dalton (1766-1844) (see section 1.1), a British chemist and physicist, developed Lavoisier's ideas further when he produced a theory about the structure of elements and compounds. Dalton proposed that:

Antoine Lavoisier

- all elements are made up of very small solid particles called atoms
- the atoms of a given element are alike and have the same weight
- the atoms of different elements are different
- chemical compounds are formed when atoms combine
- a given compound is always made up of the same number and type of atoms.

Dalton's proposals showed that many of Lavoisier's 'elements' were not in fact elements.

During the 19th century chemists were discovering and finding out the behaviour of a large number of new elements. Imagine that as each new element was discovered, the following information was recorded on a card.

Name of element ....
Chemical and physical properties
....
....
....
Atomic weight* ....

(* now called relative atomic mass)

Chemists, just like all good scientists, began to realise that it was necessary to classify this increasing amount of knowledge about the elements into some kind of pattern. The pattern linked together elements with similar properties.

In 1817 Johann Wolfgang Dobereiner, a German chemist, who would have had only a few cards available, showed that the atomic weight of strontium was almost midway between that of calcium and barium and that these three elements had similar properties. Later he showed that the cards of certain other elements with similar properties could also be arranged in sets of three (triads) – for example chlorine, bromine and iodine; iron, cobalt and manganese.

Unfortunately, other chemists failed to grasp the importance of these triads mainly because of the limited number of elements used.

In 1864 John Newlands, a British chemist, proposed and published the idea that if cards were arranged in groups of seven in order of atomic weight, then every eighth element in this grouping shared similar chemical and physical properties. For example lithium and sodium, sodium and potassium, magnesium and calcium.

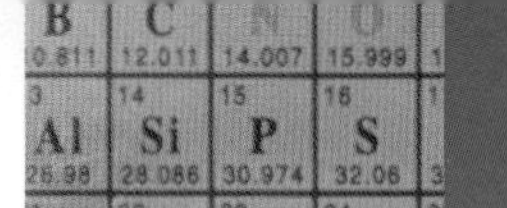

Newland's arrangement of some of the elements is shown in Figure 5.2.

Figure 5.2
Newland's arrangement of some of the elements

1	2	3	4	5	6	7
H	Li	Be	B	C	N	O
F	Na	Mg	Al	Si	P	S
Cl	K	Ca				

Dmitri Mendeleev

This idea came to be known as the 'law of octaves' because it suggested comparisons with the musical scale. But because of this comparison the law was ridiculed until the work of Mendeleev showed how much truth there was in Newland's idea.

Dmitri Mendeleev (1834–1907) was a Russian chemist who was professor at the University of St Petersburg. At that time there were no good chemistry textbooks, so Mendeleev wrote his own. It was for this book that he started to classify the elements on the basis of their known properties and atomic weights.

In 1869 his first version of a periodic table was published. This was further refined and in 1871 a second version was published. By the time he produced his classification he had information on more than 60 cards. Mendeleev arranged the elements initially in order of atomic weight, but then proceeded to try to make sure that elements with similar properties were in the same vertical column.

	Group 1		Group 2		Group 3		Group 4		Group 5		Group 6		Group 7	
Period 1	H (1)													
Period 2	Li (7)		Be (9.4)		B (11)		C (12)		N (14)		O (16)		F (19)	
Period 3	Na (23)		Mg (24)		Al (27.3)		Si (28)		P (31)		S (32)		Cl (35.5)	
Period 4	K (39)	Cu (63)	Ca (40)	Zn (65)	?	?	Ti (48)	?	V (51)	As (75)	Cr (52)	Se (78)	Mn (55)	Br (80)
Period 5	Rb (85)	Ag (108)	Sr (87)	Cd (112)	Y (88)	In (113)	Zr (90)	Sn (118)	Nb (94)	Sb (122)	Mo (96)	Te (125)	?	I (127)

Figure 5.3
Part of Mendeleev's 1869 periodic table. The numbers in brackets are the values for the atomic weights used by Mendeleev

Many scientists did not accept the table at first and treated its contents as no more than an interesting curiosity. But Mendeleev's classification contained many gaps and he believed that the gaps were for elements not yet discovered.

Mendeleev realised that he could prove the value of his classification if it could be used as a tool to predict the missing elements. To do this he set about predicting the properties of the elements likely to occupy the three spaces shaded in the table.

Mendeleev predicted that the missing element in group 4 would have an atomic weight that was likely to be the average of the atomic weights of silicon (Si) and tin (Sn). He also predicted the colour, the density, the melting point and the formula of the oxide of the missing element, which he called ekasilicon. In 1886 germanium was discovered and found to match almost completely all the predictions made for ekasilicon.

Mendeleev made further predictions for the properties of the two elements missing from group 3. The element missing between calcium (Ca) and titanium (Ti) he called ekaboron and the element missing next to zinc (Zn) he called ekaaluminium. In 1875 gallium was discovered and found to match almost completely the predictions made for ekaaluminium and in 1879 scandium was discovered and found to match almost completely the predictions made for ekaboron.

The correctness of Mendeleev's predictions proved to other scientists the importance of the ideas behind his periodic table. Many scientists began to use the table as a working tool that would help them complete the gaps in the rest of the table. Indeed in the 30 years following the publication of the periodic table many more elements were discovered.

The periodic table has undergone two main revisions since being proposed by Mendeleev. The first revision was the inclusion of a new family of elements called the inert or noble gases whose existence was unsuspected for most of the 19th century. The second revision was brought about by the work of Henry Moseley (1887–1915).

Henry Moseley was a British chemist who used X-ray spectra to study atomic structure. His work showed that the **atomic number** and not the atomic weight caused the repeated patterns of behaviour of different elements. In Mendeleev's version of the periodic table based on arranging the elements in order of their relative atomic weights, with elements with similar properties being in the same column most elements are in appropriate groups. However a few are not. Argon atoms, for example, have a greater relative atomic mass (40) than potassium atoms (39). Because of Moseley's work, the elements in the modern periodic table are arranged in order of the atomic number (proton number), and all elements are in appropriate groups.

The contents of the modern periodic table can be used as a tool to predict:

- whether an element is a metal or a non-metal
- the physical properties of an element
- the relative reactivities of the various elements
- the reactions between elements
- the charge on the ions of the elements
- chemical formulae.

All of these predictions are made possible because the modern periodic table is a comprehensive summary of the structure of the atom of each of the known elements.

The modern periodic table is shown in Figure 5.4, on the next page.

Figure 5.4
The periodic table

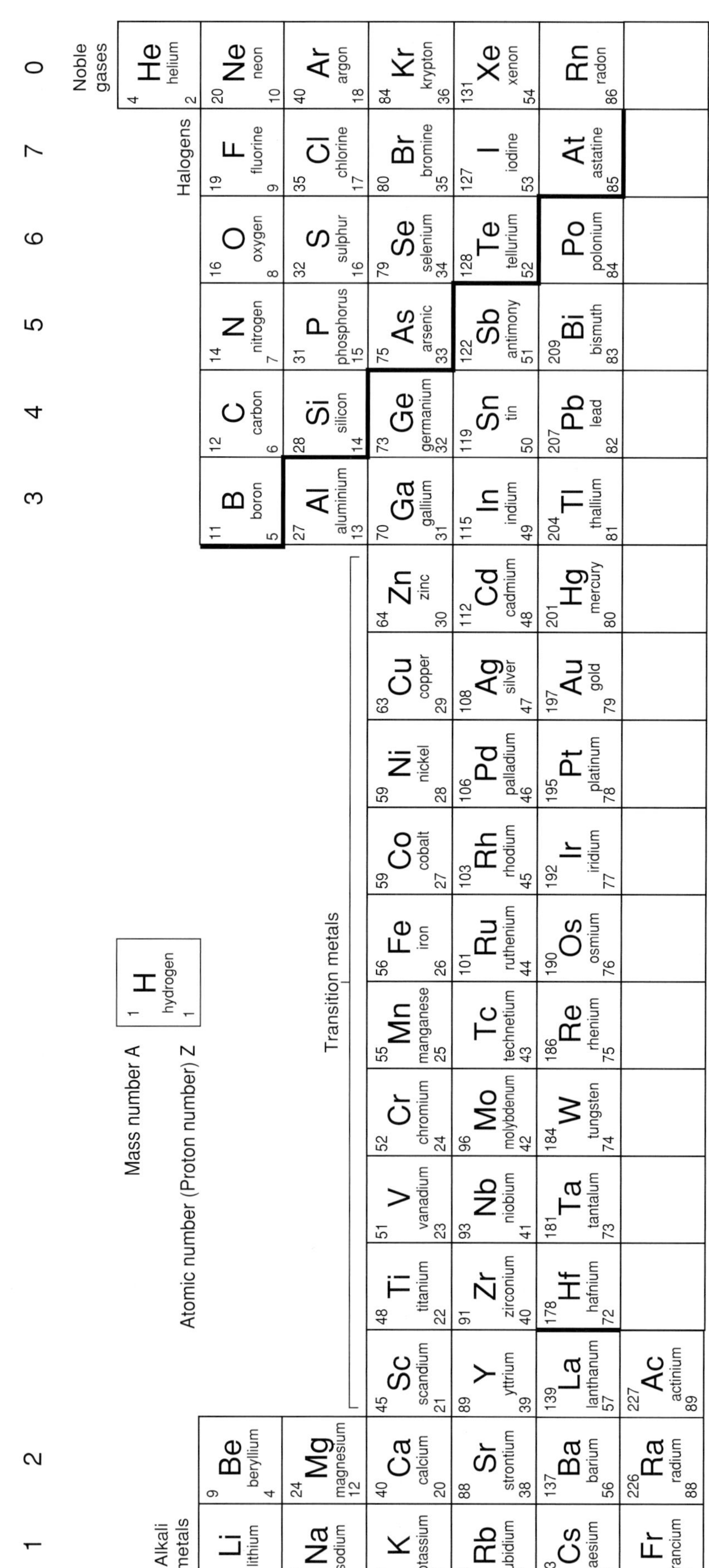

1	2											3	4	5	6	7	0
																	4 He helium 2
7 Li lithium 3	9 Be beryllium 4											11 B boron 5	12 C carbon 6	14 N nitrogen 7	16 O oxygen 8	19 F fluorine 9	20 Ne neon 10
23 Na sodium 11	24 Mg magnesium 12											27 Al aluminium 13	28 Si silicon 14	31 P phosphorus 15	32 S sulphur 16	35 Cl chlorine 17	40 Ar argon 18
39 K potassium 19	40 Ca calcium 20	45 Sc scandium 21	48 Ti titanium 22	51 V vanadium 23	52 Cr chromium 24	55 Mn manganese 25	56 Fe iron 26	59 Co cobalt 27	59 Ni nickel 28	63 Cu copper 29	64 Zn zinc 30	70 Ga gallium 31	73 Ge germanium 32	75 As arsenic 33	79 Se selenium 34	80 Br bromine 35	84 Kr krypton 36
85 Rb rubidium 37	88 Sr strontium 38	89 Y yttrium 39	91 Zr zirconium 40	93 Nb niobium 41	96 Mo molybdenum 42	Tc technetium 43	101 Ru ruthenium 44	103 Rh rhodium 45	106 Pd palladium 46	108 Ag silver 47	112 Cd cadmium 48	115 In indium 49	119 Sn tin 50	122 Sb antimony 51	128 Te tellurium 52	127 I iodine 53	131 Xe xenon 54
133 Cs caesium 55	137 Ba barium 56	139 La lanthanum 57	178 Hf hafnium 72	181 Ta tantalum 73	184 W tungsten 74	186 Re rhenium 75	190 Os osmium 76	192 Ir iridium 77	195 Pt platinum 78	197 Au gold 79	201 Hg mercury 80	204 Tl thallium 81	207 Pb lead 82	209 Bi bismuth 83	Po polonium 84	At astatine 85	Rn radon 86
Fr francium 87	226 Ra radium 88	227 Ac actinium 89															

Elements 58–71 and 90–103 have been omitted.

The value used for mass number is normally that of the commonest isotope, eg ^{35}Cl not ^{37}Cl
Bromine is approximately equal proportions of ^{74}Br and ^{81}Br

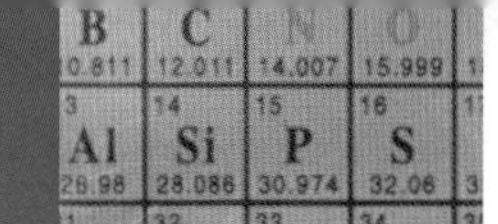

Summary

- Early attempts to organise the elements into patterns that reflected their behaviour arranged the elements in order of atomic weight. The modern **periodic table** arranges the elements in increasing order of atomic number.

Topic questions

1. During the 19th century there was a need to develop some form of classification for the elements. Why?
2. It was important that the atomic weight of each element should be measured as accurately as possible. Why?
3. Newland's table was considered to be less useful than that produced by Mendeleev. Why?
4. In what way were the findings of Newlands similar to those of Mendeleev?
5. Use the atomic weights given in Figure 5.3 to predict the likely atomic weight for the element missing from Group 4. Explain how you got to your answer.
6. Explain why Mendeleev's periodic table was finally accepted by other scientists?
7. Which family of elements was missing from Mendeleev's periodic table? Give a reason.
8. Which change to Mendeleev's periodic table resulted from the work of Moseley?

Note: When answering examination questions about the development of the periodic table you will not be expected to remember all the information provided in this section. The examination questions will provide all the background information you might need for your answer.

5.2 Patterns in the periodic table

Co-ordinated	Modular
11.11	08 (10.13)

Period

A **period** is a horizontal row of elements in the periodic table in order of increasing atomic number. Period 2 contains sodium, magnesium, aluminium, silicon, phosphorus, sulphur, chlorine and argon. A new period is started each time a new outer shell of electrons (see section 1.1) is started.

The chemical properties of the elements in a period change gradually across the period as an extra outer electron is added with each new element.

Group

A **group** is a vertical column of elements in the periodic table. The atoms of elements in the same group have the same number of electrons in their outer shell. Because of this, elements in the same group have similar chemical properties.

Each group is given a number which represents the number of electrons in the outer shell of their atoms. (In Group 0 the outer shell is full.)

Metals are found in groups on the left-hand side of the periodic table, e.g. Groups 1 and 2. The non-metals are found in groups on the right-hand side of the periodic table.

There is a gradual change from metal to non-metal as you go across a period. The metals found in the central block are called the **transition metals** or **transition elements**.

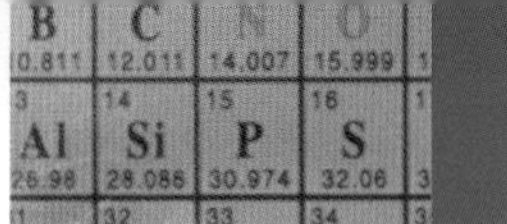

Group 0: The noble gases

Group 0 is on the far right-hand side of the periodic table. Each element has atoms with a full outer shell of electrons.

Properties of noble gases

1 They are all gases at room temperature.

2 They are all monatomic and exist as single atoms e.g. He and Ne. (Most elements which are gases form molecules containing more than one atom e.g. H_2, O_2 and O_3.)

3 They are unreactive (**inert**) towards other elements.

Noble gases are unreactive and monatomic because their atoms have a full outer shell of electrons and do not need to gain, lose or share any other electrons to become more stable.

Figure 5.5 *Physical properties of helium, neon and argon*

Element	Atomic number	Relative atomic mass	Melting point (°C)	Boiling point (°C)
helium	2	4.0	−270	−269
neon	10	20.2	−249	−246
argon	18	40.0	−189	−186

Did you know?

Originally the noble gases were called the inert gases because they were thought to be completely unreactive. However, xenon was found to react with fluorine, which is a very reactive element, and the group name had to be changed.

Uses of noble gases

- Helium is used in airships because it is much less dense than air. It provides the buoyancy needed and is much safer than hydrogen because it is non-flammable.
- Neon is used to give the red coloured light used in advertising signs.
- Argon is used as the inert gas in light bulbs. It is unreactive and, unlike air, does not react with the hot metal filament in the light bulb. This helps the light bulb to last longer.

Group 7: the halogens

Group 7 is near the right-hand side of the periodic table. Each element has atoms with seven electrons in the outer shell of electrons.

The atoms of the **halogens** join together to form diatomic molecules e.g. Cl_2.

Figure 5.6 *Physical properties of chlorine, bromine and iodine*

Element	Atomic number	Relative atomic mass	Melting point (°C)	Boiling point (°C)
chlorine	17	35.5	−101	−34
bromine	35	80	−7	58
iodine	53	127	114	183

Properties of chlorine, bromine and iodine

1 Melting points and boiling points are low, but increase down the group as the atomic number increases.

2 They are brittle/crumbly when solid.

3 They are poor conductors of heat and electricity.

4 Reactivity decreases down the group as the atomic number increases.

In a reaction with a metal, a halogen atom will gain an electron to form a halide ion, X^-. Adding an extra electron to the outer shell of electrons becomes more difficult as the size of the atom increases down the group. This is because the outer electrons are further from the nucleus and less strongly attracted. The extra inner electron shells also shield the outer electrons from the full attractive force from the nucleus. Both of these effects mean that the extra electron is more difficult to add to the outer shell of electrons as part of a reaction.

Did you know?

Astatine, At, the last member of the halogen group is very rare. It is estimated that there is less than 0.2 grams of astatine on the Earth.

5 They are all non-metals with coloured vapours.
- Fluorine is a straw coloured gas
- Chlorine is a yellow-green gas
- Bromine is a red liquid with a red-brown vapour
- Iodine is a grey shiny solid and gives off a violet vapour on heating.

6 They react with hydrogen forming covalent hydrogen **halides**.
Hydrogen reacts with chlorine to form hydrogen chloride.

$$H_2 + Cl_2 \rightarrow 2HCl$$

Hydrogen reacts with bromine to form hydrogen bromide.

$$H_2 + Br_2 \rightarrow 2HBr$$

Hydrogen reacts with iodine to form hydrogen iodide.

$$H_2 + I_2 \rightarrow 2HI$$

7 **Displacement** reactions.
Each halogen will displace a halogen lower in the group to form a solution of the lower halide.
Chlorine will displace bromine and iodine from bromide and iodide solutions.

$$Cl_2 + 2KBr \rightarrow Br_2 + KCl$$

$$Cl_2 + 2KI \rightarrow I_2 + KCl$$

Bromine will displace iodine from iodide solutions but not chlorine from chloride solutions.

$$Br_2 + 2KI \rightarrow I_2 + KBr$$

8 They react with alkali metals. Ionic solids are formed with the halogen ion, Cl^-, Br^- and I^-, having a single negative charge.

9 They react with non-metallic elements to form covalent molecular compounds e.g. CCl_4.

Summary

- A **period** is a horizontal row of elements in the periodic table arranged in order of increasing atomic number.
- A **group** is a vertical column of elements whose atoms have the same number of electrons in their outer shell.
- Similarities and differences in the properties of elements can be explained in terms of their electronic structure.
- More than three quarters of the elements are metals found in Groups 1 and 2 and in the block of transition metals.
- The elements in Groups 7 and 0 are non-metals.
- The elements in Group 7 are called the **halogens**, their ions carry a single negative charge.
- The reactivity of the halogens decreases going down the group.
- The elements in Group 0 are called the **noble gases**.
- The noble gases are unreactive because their atoms have a complete outer shell.

Topic questions

1 What is the name given to:
a) each horizontal row of elements in the periodic table?
b) each column of elements in the periodic table?

2 What happens to the atomic numbers of the elements in a particular period as you move from left to right across the periodic table?

3 What can you say about the number of electrons in the outer shell of the elements in a particular group?

4 What can you say about the outer shell of the elements in Group 0?

5 In which groups are the non-metals found?

6 a) Give three properties of the noble gases.
b) What happens to the boiling points of the noble gases as their atomic number increases?

7 a) What is the name of the elements in Group 7?
b) How many electrons are in the outer shell of the elements in Group 7?
c) What happens to the reactivity of the Group 7 elements as their atomic number increases?

8 Use your periodic table to work out the number of electrons in the outer electron shells of the following elements:
a) sodium
b) chlorine
c) oxygen
d) phosphorus
e) magnesium
f) carbon
g) aluminium
h) neon
i) fluorine
j) boron

5.3	
Co-ordinated	Modular
11.11	08 (10.13)

Metals and the periodic table

Group 1: The alkali metals

Group 1 is on the left-hand side of the periodic table. Each element has atoms with one electron in the outer shell of electrons.

Figure 5.7 *Physical properties of lithium and sodium and potassium*

Element	Atomic number	Relative atomic mass	Melting point (°C)	Boiling point (°C)	Density (g/cm^3)
lithium	3	6.9	180	1330	0.53
sodium	11	23.0	98	883	0.97
potassium	19	39.1	64	760	0.86

Properties of lithium, sodium and potassium

1 They are metals with a low melting point and a low density (they float on water).

2 Their melting and boiling points decrease as you go down the group with increasing *atomic number* (see section 1.1).

3 They react readily with cold water to form **hydrogen** and metal hydroxides which dissolve in water to give **alkaline** solutions.

lithium + water → lithium hydroxide + hydrogen
$2Li(s) + 2H_2O(l) \rightarrow 2LiOH(aq) + H_2(g)$

sodium + water → sodium hydroxide + hydrogen
$2Na(s) + 2H_2O(l) \rightarrow 2NaOH(aq) + H_2(g)$

potassium + water → potassium hydroxide + hydrogen
$2K(s) + 2H_2O(l) \rightarrow 2KOH(aq) + H_2(g)$

The reactivity increases down the group as the atomic number increases. This is shown by the increasingly vigorous reactions of lithium, sodium and potassium with water. A test for hydrogen is described on the next page.

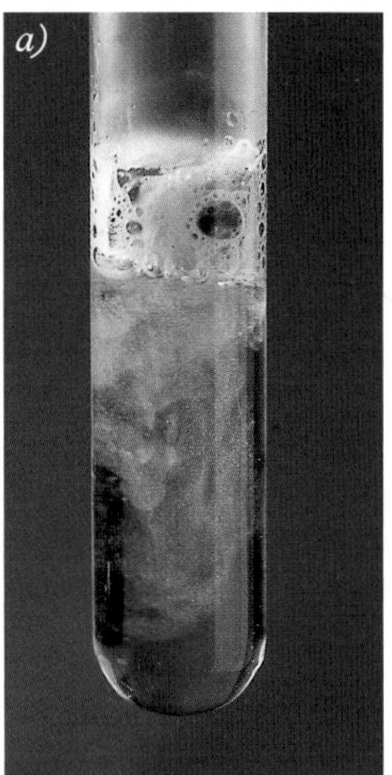

Figure 5.8
a) lithium, b) sodium and c) potassium reacting with water

In a reaction, the metal atom will lose an electron to form a metal ion, M^+. The outer electron on the atom becomes easier to remove as the size of the atom increases down the group. This is because the outer electron is further from the nucleus and less strongly attracted by the positively-charged protons. The extra inner electron shells also shield the outer electron from the full attractive force from the nucleus. Both of these effects mean that the outer electron is less firmly held by the nucleus and is easier to lose as part of a reaction.

4 They react with halogens to give ionic halides.
The resulting metal ion has a single positive charge, M^+.
Lithium reacts with chlorine to give lithium chloride, Li^+Cl^-.

$$2Li(s) + Cl_2(g) \rightarrow 2LiCl(s)$$

They react with bromine to give ionic bromides.
Sodium reacts with bromine to give sodium bromide, Na^+Br^-.

$$2Na(s) + Br_2(l) \rightarrow 2NaBr(s)$$

They react with iodine to give ionic iodides.
Potassium reacts with iodine to give potassium iodide, K^+I^-.

$$2K(s) + I_2(s) \rightarrow 2KI(s)$$

5 They react with oxygen to form ionic oxides $(M^+)_2O^{2-}$.
The resulting metal ion has a single positive charge, M^+.
Lithium reacts with oxygen to form lithium oxide, $(Li^+)_2O^{2-}$.

$$4Li(s) + O_2(g) \rightarrow 2Li_2O(s)$$

Sodium reacts with oxygen to form sodium oxide, $(Na^+)_2O^{2-}$.

$$4Na(s) + O_2(g) \rightarrow 2Na_2O(s)$$

Potassium reacts with oxygen to form potassium oxide, $(K^+)_2O^{2-}$.

$$4K(s) + O_2(g) \rightarrow 2K_2O(s)$$

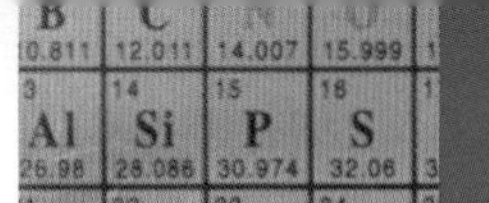

A laboratory test for hydrogen

If a lighted splint is held over the top of a test tube of hydrogen, the hydrogen burns with a squeaky explosion.

Did you know?

Sodium can be used as a coolant. Sodium is a metal with a low melting point and it readily melts to form a liquid. This liquid is an excellent heat conductor and has been used as a high temperature coolant in nuclear reactors and record-breaking car engines. Special precautions have to be taken against leakage as molten sodium is explosively reactive.

Alkali metal compounds

The compounds of alkali metals are white solids. They are soluble in water and form colourless solutions.

Alkali metal halides are ionic compounds, which dissolve in water to form separate aqueous ions. Each ion is surrounded by water molecules in solution.

$$Na^{+}Cl^{-}(s) \xrightarrow{(H_2O)} Na^{+}(aq) + Cl^{-}(aq)$$

Did you know?

The sea contains a number of dissolved ionic substances that have been brought in by rivers and remain in the sea after the water has evaporated as part of the water cycle. The commonest ionic substance is sodium chloride, or 'salt', but magnesium chloride, calcium chloride and potassium chloride are also present in large amounts.

Patterns of behaviour

Alkali metal hydroxides are ionic compounds that dissolve in water to form alkaline solutions with a pH of >7.

These alkaline solutions contain **hydroxide ions**, $OH^{-}(aq)$. Sodium hydroxide dissolves in water to form $Na^{+}(aq)$ and $OH^{-}(aq)$ ions.

$$Na^{+}OH^{-}(s) \xrightarrow{(H_2O)} Na^{+}(aq) + OH^{-}(aq)$$

Industrially, sodium hydroxide solution is formed by the electrolysis of sodium chloride solution. Chlorine and hydrogen are also formed at the same time. The simultaneous production of three useful industrial chemicals helps to make the whole process more economical.

Summary

- The elements in Group one are called the **alkali metals**, their metal ion carries a single positive charge.
- The reactivity of the alkali metals increases going down the group.
- The alkali metals react with water to form hydrogen and metal hydroxides. The metal hydroxides dissolve in water to give **alkaline** solutions.
- The test for hydrogen is that it burns with a squeaky explosion when lighted (for example with a burning splint).

Topic questions

1 a) What is the name given to the elements in Group 1?
b) How many electrons are there in the outer shell of these elements?

2 What happens to the reactivity of the Group 1 elements as their atomic number increases?

3 Write down the word equation for the reaction between sodium and water.

4 What is the laboratory test for hydrogen?

5 a) Write down the word equation for the reaction between sodium and bromine.
b) Write down the balanced symbol equation for the reaction between potassium and iodine. (Ignore state symbols)

5.4 Patterns in the transition elements

Co-ordinated	Modular
11.12	0 5 (10.14)

The **transition elements** are found between Groups 2 and 3 in the periodic table.

Element	Atomic number	Relative atomic mass	Melting point (°C)	Density (g/cm^3)
manganese	25	54.9	1244	7.2
iron	26	55.8	1535	7.9
copper	29	63.5	1083	8.9
zinc	30	65.4	420	7.1
mercury	80	201.0	−39	13.6

Figure 5.9 *Physical properties of manganese, iron, copper, zinc and mercury*

Properties of the transition elements

1 They are metals with high melting points and high densities (except for mercury).

2 They can be used as catalysts to speed up reactions. For example, iron is used as a catalyst in the Haber process to make ammonia from nitrogen and hydrogen (see section 3.2).

3 They are hard, tough and strong.

4 They are much less reactive than the alkali metals and so do not corrode so quickly with oxygen and/or water.

Did you know?

Manganese, iron, copper and zinc are needed by plants and animals in very small amounts for healthy growth – they are known as trace elements. Iron is found in red blood cells as part of a haemoglobin molecule, which carries oxygen from the lungs to all body cells. A lack of iron in the body will result in anaemia.

Properties of compounds

1 They form coloured compounds. These can often be seen in pottery glazes of various colours or in weathered copper (green).

2 They have catalytic properties and speed up some reactions e.g. manganese (IV) oxide, MnO_2, helps to decompose hydrogen peroxide into water and oxygen.

Uses

- Cast iron is used to make manhole covers.
- Iron is used as a catalyst to manufacture ammonia.
- Copper is used in electrical wiring and domestic hot water pipes.
- Zinc is used in **galvanising** iron to prevent it from rusting.

Summary

- The **transition elements** all have metallic properties and form coloured compounds.
- Many transition elements are used as catalysts.

Topic questions

1 Identify the following elements from their descriptions:
   a) a metal that is used in household electrical wiring.
   b) a metal that is used as a catalyst in the manufacture of ammonia.
   c) a metal that is used to galvanise iron.
   d) a metal oxide that acts as a catalyst.

2 Between which groups in the periodic table are the transition elements found?

3 Why are these elements often called the transition metals?

4 Give four properties of transition elements.

5 Give two properties of the compounds formed from transition elements.

5.5 Patterns in the reactions of metal halides (halogens)

Co-ordinated	Modular
11.12	0 8 (10.14)

Common salt (sodium chloride, NaCl)

Most of the salt used on our foods comes not from the sea but from mines in Cheshire. Because salt is soluble in water it is extracted from deep underground by a process called solution mining. In hot countries salt is obtained from sea water by evaporation.

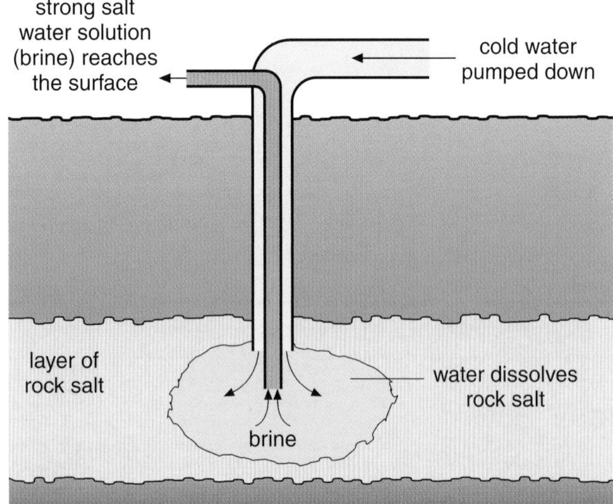

Figure 5.10
Solution mining

Figure 5.11
Salt pans – the water evaporates and the salt remains

Sodium chloride is a compound made from sodium (a very reactive alkali metal) and chlorine (a very poisonous and reactive halogen). This illustrates the fact that the chemical properties of the separate elements in a compound are completely different from the properties of the compound they form.

The electrolysis of sodium chloride solution (brine)

The **electrolysis** of brine is an important industrial process. It is important because of the three very useful chemicals produced, chlorine, sodium hydroxide and hydrogen.

Electrolysis of sodium chloride solution
Hydrogen (H_2)
Used in the making of margarine and ammonia
Chlorine (Cl)
Used to kill bacteria in drinking water and in swimming baths. Used in the making of disinfectants, hydrochloric acid, bleach and the plastic polymer PVC
Sodium hydroxide (NaOH)
Used in the making of soaps, detergents, paper and ceramics

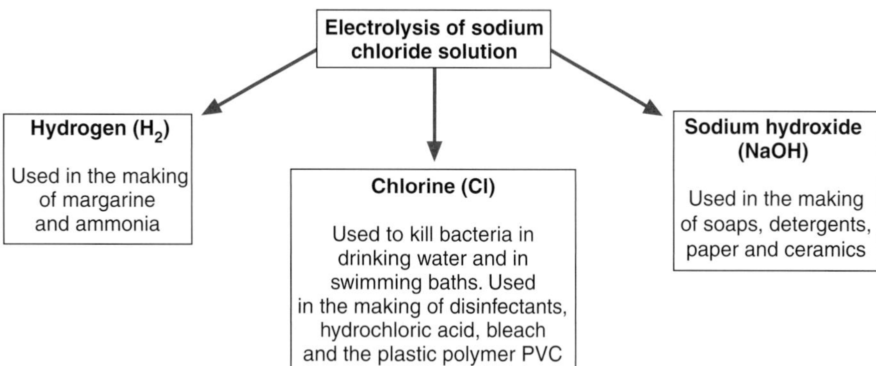

Figure 5.12
Products of electrolysis of sodium chloride

What happens during the electrolysis of sodium chloride solution?

Because sodium chloride solution is the conducting solution (**electrolyte**) the ions present in it are H^+ ions and OH^- ions from the water and Na^+ ions and Cl^- ions from the sodium chloride.
The **anode** has a positive charge because it has lost electrons to the cathode. The anode therefore can gain electrons.
The OH^- and Cl^- ions are negatively charged so have gained electrons.
These negative ions are attracted to the positive electrode, the anode, where they can lose the extra electrons to make them into atoms. However, only the chloride ions become atoms. These atoms pair up to form Cl_2 molecules. The hydroxide ions remain in the solution. So, at the anode chlorine gas is released.
The reaction at the anode can be summarised as:
chloride ions – electrons → chlorine
$2Cl^-$ (l) – $2e^-$ → Cl_2 (g)
This is an example of oxidation.
← OH^-
H^+ →
← Cl^-
H^+ →
The **cathode** has a negative charge because it has gained extra electrons from the anode. The cathode has electrons to lose.
The H^+ and Na^+ ions are positively charged so have lost electrons (see section 4.2).
These positive ions are attracted to the negative electrode, the cathode, where they can collect the extra electrons to make them into atoms.
However, because sodium is more reactive than hydrogen the sodium ions remain as ions in the solution and only the hydrogen ions form atoms. These atoms pair up to form H_2 molecules. So, at the cathode hydrogen gas is released.
The reaction at the cathode can be summarised as:
hydrogen ions + electrons → hydrogen
$2H^+$ (aq) + $2e^-$ → H_2 (g)
This is an example of reduction.
The Na^+ ions and the OH^- ions remain in solution to form a solution of sodium hydroxide

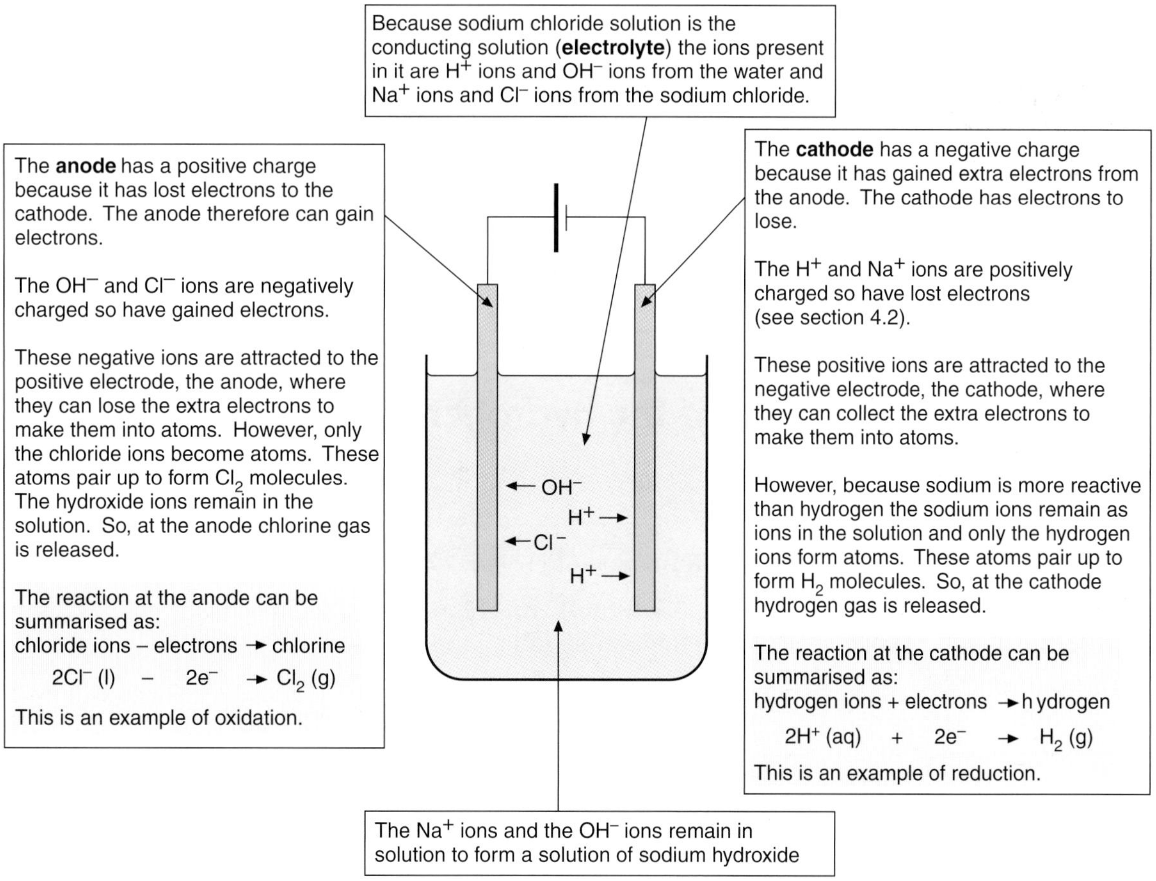

Figure 5.13
Electrolysis of sodium chloride solution

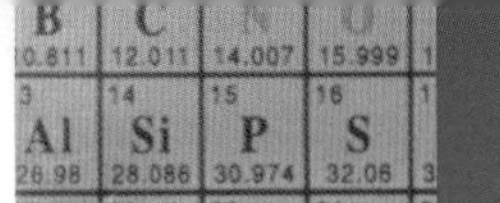

A health warning about chlorine

Chlorine gas even in small amounts is poisonous. If breathed in, it affects the cells lining the lungs, making them produce large quantities of fluid. If a lot of chlorine gas is breathed in then the large amounts of fluid produced can fill the lungs and can cause permanent lung damage or even death.

Did you know?

Chlorine was used as a poison gas in the first World War. When the wind was in the right direction, it was released towards the opposing army. It is denser than air and rolled along the ground until it reached the enemy trenches, filling them with the choking gas. However, if the wind changed direction, the chlorine gas could blow back towards the attacking army.

The laboratory test for chlorine gas

When chlorine is used as a bleach it combines with the coloured dyes in a cloth and turns them into colourless compounds. This property is made use of in the laboratory test for chlorine. In the test a piece of damp litmus paper is held above a small amount of the gas. If the gas is chlorine the litmus paper loses its colour.

The effect of light on silver halides (halogens)

Almost all types of photograph rely on the light-sensitive properties of the silver halides silver chloride, silver bromide and silver iodide. Photographic film consists of a thin layer of gelatin on a base of transparent plastic. The emulsion contains crystals of silver halides. When the film is exposed to light the silver halide crystals undergo a chemical change. As the film is developed the silver halide crystals in those areas exposed to most light become reduced to particles of metallic silver. Photographic paper works in much the same way as the film, but fewer silver halide crystals are used.

X-rays and radiation from radioactive substances produce the same changes to silver halide crystals as light.

Reactions of the hydrogen halides

Hydrogen halides are all gases that form acidic solutions when added to water.

Did you know?

Hydrofluoric acid is so corrosive that it will corrode glass. It is stored in lead, steel or plastic containers.

This is due to the formation of $H^+(aq)$ ions.

Hydrogen chloride gas forms hydrochloric acid when it is dissolved in water. Hydrogen chloride in water gives H^+ and Cl^- ions.

$$HCl(g) \xrightarrow{(H_2O)} H^+(aq) + Cl^-(aq)$$

Hydrogen reacts with the halogens in ways that reflect the relative reactivities of the Group 7 elements.

Figure 5.4 *Reactions of hydrogen and halogens*

Rate of reaction	Action with water
A mixture of hydrogen and fluorine will react explosively to form hydrogen fluoride (HF)	Forms hydrofluoric acid
Chlorine and hydrogen can be mixed together without a reaction if they are kept in the dark. If the mixture is exposed to sunlight they react together explosively to form hydrogen chloride (HCl)	Forms hydrochloric acid (a commonly used laboratory acid)
A mixture of hydrogen and bromine needs to be heated if they are to react and form hydrogen bromide (HBr)	Forms hydrobromic acid
Very little reaction between hydrogen and iodine even if they are heated	

Summary

- The electrolysis of sodium chloride solution produces chlorine gas at the anode, hydrogen gas at the cathode and a solution of sodium hydroxide.
- The test for chlorine is that it will decolorise moist litmus paper.
- Silver halides react in light to form particles of silver.

Topic questions

1 Explain why salt can be mined using a process called 'solution mining' but coal cannot.

2 What three important products are formed when a solution of sodium chloride is electrolysed? Give two uses for each product.

3 During the electrolysis of sodium chloride solution.
a) What substance is produced at the cathode?
b) What substance is produced at the anode?

Explain in terms of ions and electrons why this substance is formed.

4 How would you test a gas to find out if it was chlorine?

5 What happens to crystals of silver bromide when they are exposed to the light?

5.6	
Co-ordinated	Modular
11.12	05 (10.14)

Patterns in making metal compounds

Acids

Pure **acids** are covalent molecules (see section 1.2). They only behave as acids in water.

- A dilute acid is formed when an acid is dissolved in a lot of water.
- A concentrated acid is formed when an acid is dissolved in only a small amount of water.

Common acids used in the laboratory are shown in Figure 5.15.

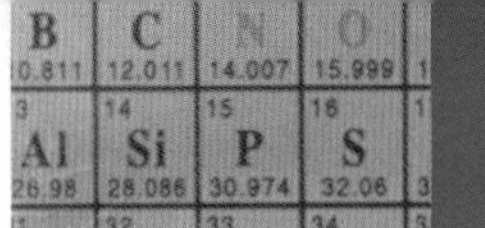

Figure 5.15 *Common laboratory acids*

Name	Formula
hydrochloric acid	HCl
nitric acid	HNO_3
sulphuric acid	H_2SO_4
ethanoic acid	CH_3COOH

Did you know?

There are may acids that occur naturally. Citric acid is found in oranges and lemons. Malic acid is found in apples. Ethanoic acid (acetic acid) is the main constituent of vinegar. Lactic acid is found in sour milk, yoghurt and some cheeses. Ants defend themselves when attacked using a spray of methanoic acid (formic acid). Stinging nettles have hollow hairs containing methanoic acid that break off in the skin and cause irritation.

Bases

Bases are usually oxides or hydroxides of metals. **Alkalis** are soluble bases.

Common bases used in the laboratory are shown in Figure 5.16.

Figure 5.16 *Common laboratory bases*

Slightly soluble bases		Soluble bases (alkalis)	
calcium hydroxide	$Ca(OH)_2$	sodium hydroxide	NaOH
calcium oxide	CaO	potassium hydroxide	KOH
		aqueous ammonia	$NH_3(aq)$

The pH scale

The **pH scale** is used to measure the acidity or alkalinity of an aqueous solution. It is a set of numbers which run from 0 to 14.

Indicators and the pH scale

Indicators are dyes which change colour when mixed with acids and alkalis. **Universal indicator** is a mixture of dyes which have been chosen to give the same order of colours as the visible spectrum (rainbow).

- An acid solution has a pH less than 7, i.e. pH 1 → 6.
- A neutral solution has a pH of exactly 7.
- An alkaline solution has a pH greater than 7, i.e. pH 8 → 14.

Universal indicator is used to measure the pH of a solution to show whether the solution is acidic, **neutral** or alkaline.

Figure 5.17 *Universal indicator colours at different pH values*

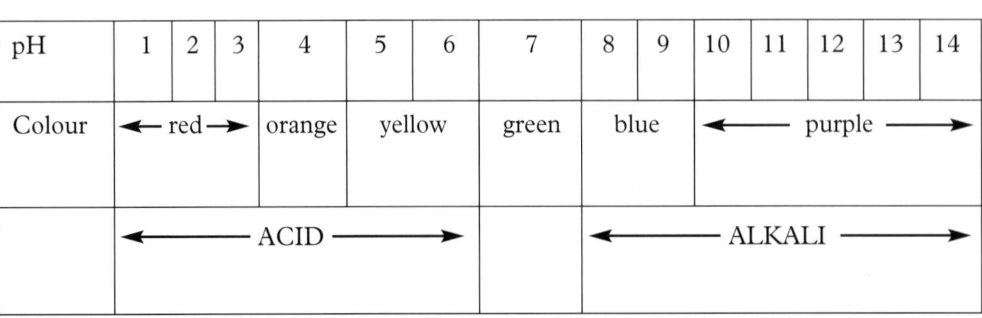

pH	1	2	3	4	5	6	7	8	9	10	11	12	13	14
Colour	← red →			orange	yellow		green	blue		← purple →				
	← ACID →							← ALKALI →						

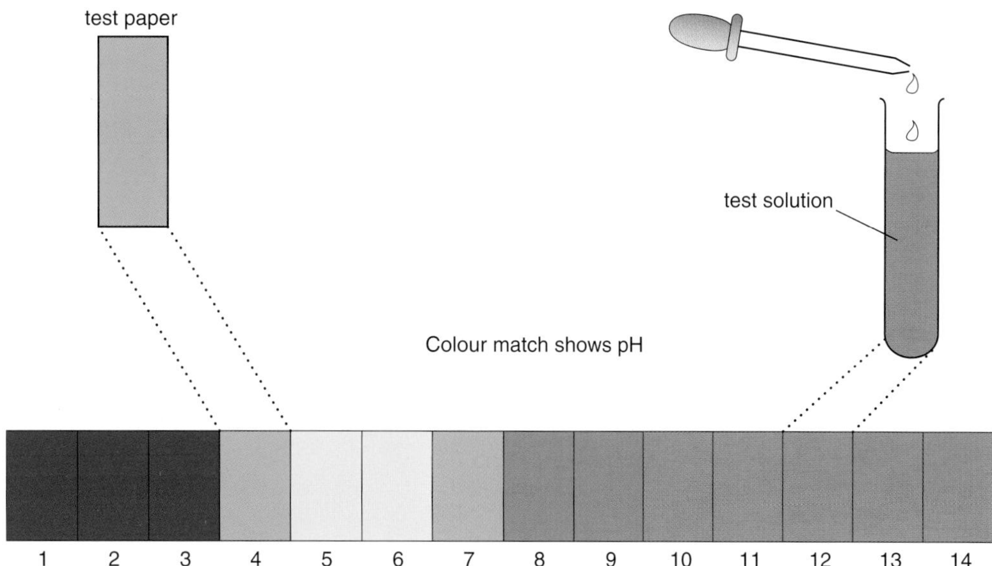

Figure 5.18 *Colour matching to the pH chart*

The solution can be tested by adding a few drops of the indicator liquid or by using indicator paper. In both cases the colour seen is compared to a coloured chart and the pH is read off.

The pH scale is a measure of the concentration of hydrogen ions (H^+) in the solution.

Did you know?

Different types of water have slightly different pH values. Rainwater has a pH of 6 due to dissolved acidic gases such as carbon dioxide and sulphur dioxide. Pure distilled water has a pH of 7. Mains water has a pH of 7.8 so that it does not react with the water pipes in a house. Sea water has a pH of 8 due to the many dissolved substances that it contains.

Acids and alkalis and ions

Acids

The formula of all acids contains hydrogen (H). An acid is a substance, which forms **hydrogen ions**, $H^+(aq)$ **ions**, when added to water.

So, HCl in water gives H^+ and Cl^-

$$HCl(g) \xrightarrow{(H_2O)} H^+(aq) + Cl^-(aq)$$

HNO_3 in water gives H^+ and NO_3^-

$$HNO_3(l) \xrightarrow{(H_2O)} H^+(aq) + NO_3^-(aq)$$

H_2SO_4 ionises into $2H^+$ and SO_4^{2-}

$$H_2SO_4(l) \xrightarrow{(H_2O)} 2H^+(aq) + SO_4^{2-}(aq)$$

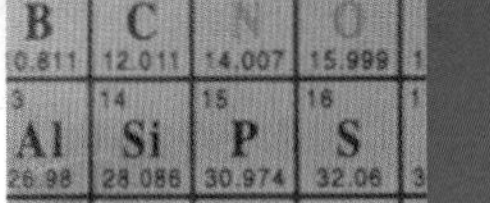

Alkalis (soluble bases)

The formula of all alkalis contains hydroxide (OH). Soluble bases (alkalis) are substances which form **hydroxide ions**, $OH^-(aq)$, when added to water. Both NaOH and KOH form hydroxide ions when dissolved in water.

NaOH in water gives

$$NaOH(s) \xrightarrow{(H_2O)} Na^+(aq) + OH^-(aq)$$

KOH in water gives

$$KOH(s) \xrightarrow{(H_2O)} K^+(aq) + OH^-(aq)$$

Aqueous ammonia reacts with water to give NH_4^+ and OH^-.

$$NH_3(g) + H_2O(l) \rightleftharpoons NH_4^+(aq) + OH^-(aq)$$

The 'double-headed' arrow, $\rightleftharpoons$, shows that the reaction is reversible (see section 8.3).

Aqueous ammonia is a base because as it dissolves in water, it also reacts with water to form OH^- ions.

Neutralisation

Neutralisation is a reaction between an acidic solution and an alkaline solution. A typical neutralisation reaction would be the reaction between the acid, hydrochloric acid, and the alkali, sodium hydroxide (see section 2.2).

The process of neutralisation can be followed using the change in colour of Universal Indicator.

$$\begin{array}{ccccccc} \text{sodium hydroxide} & + & \text{hydrochloric acid} & \rightarrow & \text{sodium chloride} & + & \text{water} \\ NaOH(aq) & + & HCl(aq) & \rightarrow & NaCl(aq) & + & H_2O(l) \end{array}$$

The solution of the sodium hydroxide is carefully added in small amounts to the hydrochloric acid and the resulting solution is tested with universal indicator paper. When the universal indicator paper turns green, the solution is neutral and no more sodium hydroxide solution need be added.

Neutralisation can be summarised as:

$$\text{alkali} + \text{acid} \rightarrow \text{salt} + \text{water}$$

The name of the salt depends upon the metal in the alkali and the acid used as can be seen from the table:

Figure 5.19 *Names of salts produced by neutralisation*

Name of alkali	Name of salt produced with Hydrochloric acid	Name of salt produced with Sulphuric acid
sodium hydroxide	sodium chloride	sodium sulphate
calcium hydroxide	calcium chloride	calcium sulphate
magnesium hydroxide	magnesium chloride	magnesium sulphate

Neutralising nitric acid produces salts called nitrates. So if ammonia solution is neutralised by nitric acid a salt called ammonium nitrate is produced.

Examples of neutralisation

- Occasionally the stomach may produce too much acid and the painful problem of acid indigestion may be felt. Indigestion tablets are a base and will neutralise the excess acid being formed by the stomach.
- Farmers add calcium hydroxide or calcium oxide to their soils in order to make them less acidic. This allows a greater range of crops to be cultivated because some plants will not grow properly on soils that are too acidic.

Neutralisation as an ionic process

When neutralisation reactions are looked at in detail it is found that all the different reactions involving acids and alkalis have the same essential ionic reaction.

The reaction between hydrochloric acid and sodium hydroxide involves the following ions.

$$H^+(aq) + Cl^-(aq) + Na^+(aq) + OH^-(aq) \rightarrow Na^+(aq) + Cl^-(aq) + H_2O(l)$$

When the spectator ions (see section 2.2) that are common to each side, $Na^+(aq)$ and $Cl^-(aq)$, are removed, only the ionic reaction between $H^+(aq)$ and $OH^-(aq)$ to make $H_2O(l)$ remains.

The essential ionic reaction for all reactions between acids and alkalis is

$$H^+(aq) + OH^-(aq) \rightarrow H_2O(l)$$

Making salts of transition metals

The oxides and hydroxides of transition metals do not dissolve in water – they are examples of insoluble bases. To make a solution of a transition metal salt, the metal oxide (or metal hydroxide) is added to an acid until no more dissolves. The excess metal oxide (or hydroxide) is removed by filtering.

Figure 5.20 *Making copper(II) sulphate*

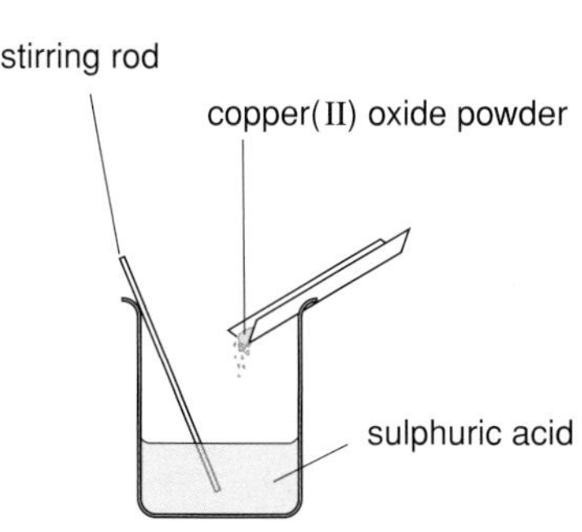

Stage 1

The copper(II) oxide is added to the acid until no more dissolves.

Stage 2

The mixture from the beaker is filtered. Collecting in the empty beaker will be a solution of copper(II) sulphate.

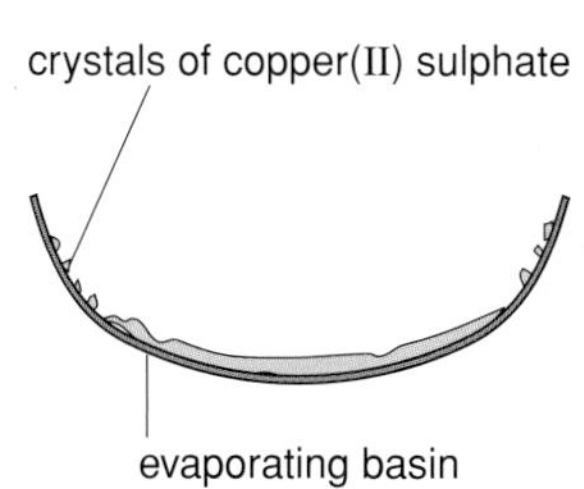

Stage 3

The copper(II) sulphate solution is left so that the water from the solution can evaporate to leave crystals of copper(II) sulphate.

The reaction can be summarised as :

copper(II) oxide + sulphuric acid → copper(II) sulphate + water

$CuO(s) + H_2SO_4(aq) \rightarrow CuSO_4(aq) + H_2O(l)$

Reactions of hydrochloric acid

Hydrochloric acid and some metal hydroxides

Metal hydroxides react with hydrochloric acid to give metal chlorides and water.

sodium hydroxide + hydrochloric acid → sodium chloride + water

$NaOH(aq) + HCl(aq) \rightarrow NaCl(aq) + H_2O(l)$

calcium hydroxide + hydrochloric acid → calcium chloride + water

$Ca(OH)_2(aq) + 2HCl(aq) \rightarrow 2CaCl_2(aq) + 2H_2O(l)$

Hydrochloric acid and ammonia

Ammonia reacts with hydrochloric acid to give a salt called ammonium chloride.

ammonia + hydrochloric acid → ammonium chloride

$NH_3 + HCl \rightarrow NH_4Cl$

Reactions of sulphuric acid

Sulphuric acid and some metal hydroxides

Metal hydroxides react with sulphuric acid to form metal sulphates and water.

sodium hydroxide + sulphuric acid → sodium sulphate + water

$2NaOH(aq) + H_2SO_4(aq) \rightarrow Na_2SO_4(aq) + 2H_2O(l)$

magnesium hydroxide + sulphuric acid → magnesium sulphate + water

$Mg(OH)_2(aq) + H_2SO_4(aq) \rightarrow MgSO_4(aq) + 2H_2O(l)$

Sulphuric acid and ammonia

Ammonia reacts with sulphuric acid to form the salt ammonium sulphate.

ammonia + sulphuric acid → ammonium sulphate

$2NH_3 + H_2SO_4 \rightarrow (NH_4)_2SO_4$

Summary

- Hydrogen halides dissolve in water to produce acidic solutions.
- The **neutralisation** reaction between an acid and an alkaline hydroxide solution produces a salt and water.
- Neutralisation can be represented as:
  $H^+ (aq) + OH^-(aq) \rightarrow H_2O\ (l)$
- Hydrochloric acid produces chlorides.
- Sulphuric acid produces sulphates.
- Nitric acid produces nitrates.
- **Hydrogen ions** H^+ make solutions acidic.
- **Hydroxide ions** OH^- make solutions alkaline.

Topic questions

1 What type of solution would have:
  a) a pH range of 1 to 6
  b) a pH of 7
  c) a pH range of 8 to 14?

2 What element do all acids contain?

3 a) What are the oxides or hydroxides of metals called?
  b) What are soluble bases called?

4 a) Write down the word equation for the reaction between sodium hydroxide solution and hydrochloric acid.

b) Write down the balanced symbol equation for the reaction between sodium hydroxide solution and hydrochloric acid. Include state symbols.
c) What is the chemical name of the salt produced in this reaction?

5 a) Write down the word equation for the reaction between copper(II) oxide and sulphuric acid.
b) Write down the balanced symbol equation for this reaction. Include state symbols.
c) What is the chemical name of the salt produced in this reaction?

6 a) Write down the word equation for the reaction between calcium hydroxide solution and hydrochloric acid.
b) Write down the balanced symbol equation for this reaction. Include state symbols.
c) What is the chemical name of the salt produced in this reaction?

7 a) Write down the word equation for the reaction between ammonia and hydrochloric acid.
b) Write down the balanced symbol equation for this reaction. Ignore state symbols.
c) What is the chemical name of the salt produced in this reaction?

8 a) Write down the ion present in all acids.
b) Write down the ion present in all alkalis.
c) Write down the ionic equation that represents neutralisation. Include state symbols.

Examination questions

1 Part of the periodic table which Mendeleev published in 1869 is shown below.

	Group 1	Group 2	Group 3	Group 4	Group 5	Group 6	Group 7
Period 1	H						
Period 2	Li	Be	B	C	N	O	F
Period 3	Na	Mg	Al	Si	P	S	Cl
Period 4	K Cu	Ca Zn	* *	Ti *	V As	Cr Se	Mn Br
Period 5	Rb Ag	Sr Cd	Y In	Zr Sn	Nb Sb	Mo Te	* I

a) Name **two** elements in Group 1 of Mendeleev's periodic table which are **not** found in Group 1 of the modern periodic table. *(2 marks)*
b) Which group of elements in the modern Periodic Table is missing on Mendeleev's table? *(1 mark)*
c) Mendeleev left several gaps on his periodic table. These gaps are shown as asterisks(*) on the table above.
Suggest why Mendeleev left these gaps. *(1 mark)*
d) Complete the following sentence.
In the **modern** periodic table the elements are arranged in the order of their ______________ numbers. *(1 mark)*

2 Part of the periodic table is shown below. Use the information to help you answer the questions which follow.

H							He
Li	Be	B	C	N	O	F	Ne
Na	Mg	Al	Si	P	S	Cl	Ar

a) Write the symbol for;
i) chlorine; *(1 mark)*
ii) sodium; *(1 mark)*
b) i) What is the symbol of the element which is in Group 2 and Period 3? *(1 mark)*
ii) What name is given to Group 7? *(2 marks)*
c) The arrangement of electrons in sulphur (S) is 2.8.6. Write the arranagement of electrons for:
i) neon (Ne) *(1 mark)*
ii) aluminium (Al) *(1 mark)*
d) The periodic table is an arrangement of elements in order of increasing atomic number. What is the atomic number of an element? *(1 mark)*
e) What is the name of the uncharged particle in the nucleus of an atom? *(1 mark)*

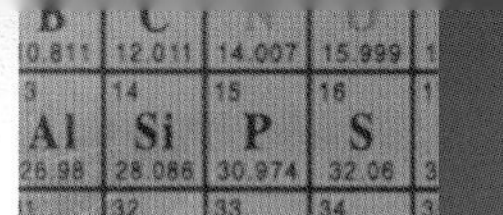

3 Chlorine, hydrogen and sodium hydroxide are produced by the electrolysis of sodium chloride solution.

A student passed electricity through sodium chloride solution using the apparatus shown in the diagram.

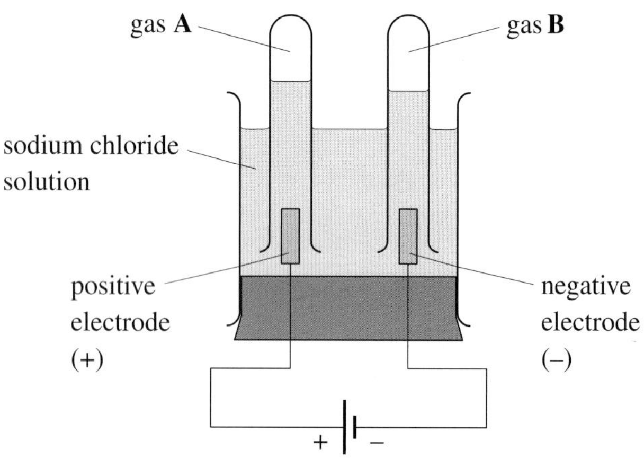

a) Name gas A and gas B *(1 mark)*
b) Describe and give the result of a test you could do in a school laboratory to find out which gas is chlorine. *(2 marks)*
c) Chlorine is used for treating water for drinking and in swimming pools. Why? *(1 mark)*

d) i) Balance the half equation for the production of hydrogen at the electrode.

$$\_\_\_\_ H^+ + \_\_\_\_ e^- \rightarrow H_2$$ *(1 mark)*

ii) Which word, from the list, best describes the reaction in part (d)(i)?

decomposition **cracking** **neutralisation** **oxidation** **reduction**

4 Acids and bases are commonly found around the home.

a) Baking powder contains sodium hydrogencarbonate mixed with an acid. When an acid is added, the baking powder releases carbon dioxide. How could you test the gas to show that it is carbon dioxide? *(2 marks)*

b) Indigestion tablets contain bases which cure indigestion by neutralising excess stomach acids.

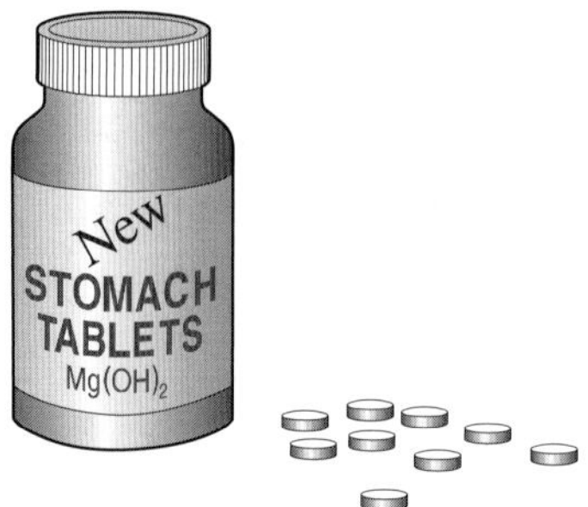

i) One type of indigestion tablet contains magnesium hydroxide. This base neutralises stomach acid as shown by the balanced chemical equation.

$$Mg(OH)_2 + 2HCl \rightarrow MgCl_2 + 2H_2O$$

Write a balanced **ionic** equation for the neutralisation reaction. *(2 marks)*

ii) How does the pH in the stomach change after taking the tablets? *(1 mark)*

5 The alkali metals, in Group 1, are the most reactive metals. The symbols and atomic numbers for the first three alkali metals are shown here.

$${}_{3}Li \quad {}_{11}Na \quad {}_{19}K$$

a) Describe what you see when sodium reacts with water. *(3 marks)*
b) Write a balanced chemical equation for the reaction of sodium with water. *(3 marks)*
c) If the reaction is repeated using lithium it is much slower, but with potassium it is much faster. Give reasons for the similarities and the differences in the reactions of lithium, potassium and sodium with water. *(4 marks)*

Chapter 6
Chemistry in action

Key terms

activation energy • **anhydrous** • **catalyst** • **denaturing** • **endothermic reaction** • **energy level diagram** • **enzyme** • **equilibrium** • **exothermic reaction** • **fermentation** • **products** • **reactants** • **reversible reaction** • **substrate**

6.1 Energy transfers in chemical reactions

Co-ordinated	Modular
11.16	07 (10.20)

Chemical reactions make new substances. In a chemical reaction the starting materials – called **reactants** – are changed into new substances – called **products**. The atoms in the reactants are re-arranged when they change into products (see section 2.2). This only happens when the reactant particles collide with each other. Sometimes when reactants are mixed there is an instant reaction, for example when magnesium is added to dilute acid the fizzing starts immediately. In other reactions the mixture of reactants has to be 'encouraged' to react by applying energy, often in the form of heat. An example of this is when natural gas (methane) and air are mixed as a gas tap is turned on. At first nothing happens – the reaction only begins when a flame is applied.

Each of these reactions get hot. This is because the reaction gives out energy. Reactions that give out energy are called **exothermic reactions**. Exothermic reactions give out energy as heat (see section 2.3).

Some reactions need energy to make them work. These reactions absorb energy. Reactions that absorb energy are called **endothermic reactions** (see section 2.3). An example of an endothermic reaction is the heating of limestone (calcium carbonate) to produce quicklime (calcium oxide) and carbon dioxide (see section 4.3).

$$CaCO_3(s) \rightarrow CaO(s) + CO_2(g)$$

calcium carbonate (limestone) → calcium oxide (quicklime)

In all chemical reactions two major changes take place:

1 The starting materials have to be decomposed by breaking the chemical bonds in them to release the elements so that they can recombine in a different way to make the products (see section 1.2). This process absorbs energy from the surroundings and is called an endothermic process.

2 Once the reactants have been decomposed, new chemical bonds are formed between the elements that have been released so that new chemical products can be produced. This releases energy to the surroundings and is called an exothermic process.

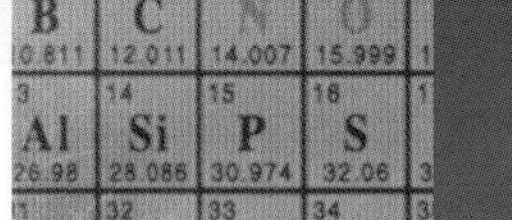

The energy changes that take place during a chemical reaction can be shown in **energy level diagrams.**

Figure 6.1 is an energy level diagram showing how the energy of the system changes during an exothermic reaction.

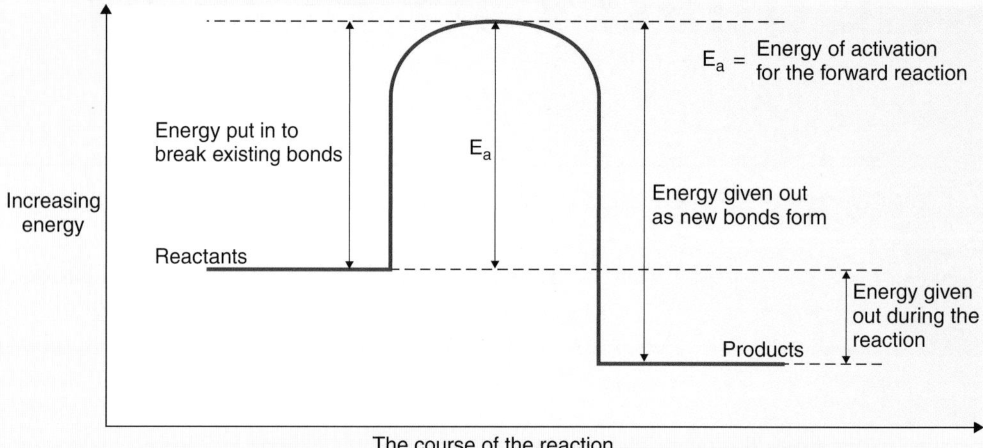

Figure 6.1
An energy level diagram for an exothermic reaction

In this case the amount of energy released as the new chemical bonds are formed is greater than the energy that is put in to break the existing bonds and start the reaction. This means there is an energy bonus and the reaction mixture will become warmer. Energy will be given out to the surroundings. Most of the energy released in this way will be heat energy but sometimes light energy is also emitted.

Figure 6.2 shows how the energy of the system changes during an endothermic reaction.

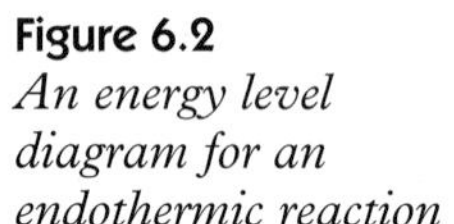

Figure 6.2
An energy level diagram for an endothermic reaction

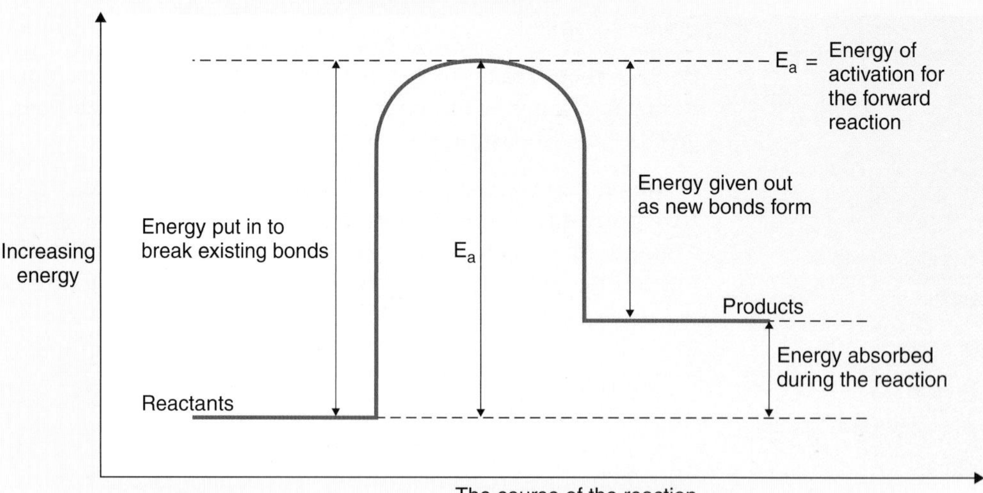

In this case the amount of energy needed to break the existing chemical bonds and start the reaction is greater than that released when new bonds are formed so there is an energy shortage. This means the reaction mixture will need to absorb energy. In some reactions this will cause the reaction mixture to become cooler. Usually endothermic reactions only work if they are heated. Endothermic reactions are less common than exothermic reactions.

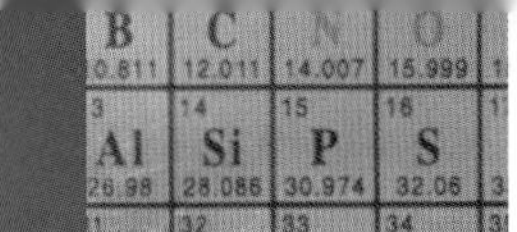

Using bond energy values to find out if a reaction is exothermic or endothermic

The table shows the energy transfers when some common chemical bonds are made or broken.

Chemical bond	Energy transferred in kJ/mol*
C—H	413
O=O	497
C=O	745
O—H	464

(*See section 7.2)

Calculating the energy transferred when methane burns in air.

The chemical equation for the reaction is: $CH_4(g) + 2O_2(g) \rightarrow CO_2(g) + 2H_2O(l)$

The bonds involved are:

$$\begin{array}{c}\text{H}\\ | \\ \text{H—C—H} \\ | \\ \text{H}\end{array} + 2 \times (\text{O=O}) \rightarrow \text{O=C=O} + 2 \times (\text{H—O—H})$$

In terms of bonds there are $4 \times (\text{C—H}) + 2 \times (\text{O=O}) \rightarrow 2 \times (\text{C=O}) + 4 \times (\text{O—H})$

The contents of the table show the bond energies involved (in kJ/mol).

Bond breaking			Bond formation		
No.	Type	Energy in	No.	Type	Energy out
4	C—H	$4 \times 413 = 1652$	2	C=O	$2 \times 745 = 1490$
2	O=O	$2 \times 497 = 994$	4	O—H	$4 \times 464 = 1856$
Total energy in = 2646 kJ/mol			Total energy out = 3346 kJ/mol		

This shows that there is a net energy transfer of heat energy to the surroundings of (3346 − 2646) kJ. The reaction is therefore exothermic, releasing 700 kJ of energy.

Activation energy

During any chemical reaction the reacting particles must first collide with each other, however, collision alone does not guarantee that the reaction will start. Reaction will only occur if the colliding particles have sufficient combined energy to get over the energy barrier between reactants and products. This minimum amount of energy that is required is called the **activation energy** – shown as E_a in Figures 6.1 and 6.2.

The energy level diagrams are similar to a fairground roller coaster – if the car carrying the passengers does not have enough energy to get up and over the first hill, the ride will not even start. This explains why some reactions occur spontaneously (i.e. as soon as the reactants are mixed), such as the reaction between magnesium and acid, and others require 'help'. A spontaneous reaction gets enough energy from the reactants and their surroundings to overcome the energy barrier between reactants and products. Reactions that require 'help' do not.

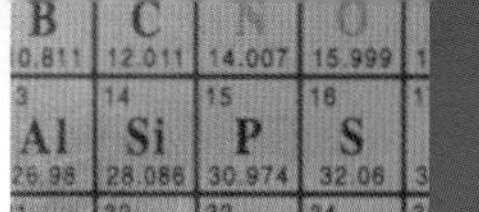

Summary

- Chemical reactions involve the re-arrangement of existing atoms to make new substances.
- Matter is neither created nor destroyed.
- Chemical bonds have to be broken before new substances can be made and this is an **endothermic** process.
- Endothermic processes require an input of energy.
- When new bonds are made, energy is released and this is an **exothermic** process.
- Exothermic processes release energy to the surroundings, usually as heat.
- The **activation energy** is the minimum amount of energy required by colliding particles before reaction can take place.
- **Energy level diagrams** link the making and breaking of bonds together to show whether the whole reaction is endothermic or exothermic.

Topic questions

1 What energy transfers take place in:
   a) exothermic reactions
   b) endothermic reactions
   c) bond breaking
   d) bond formation?

2 Explain in terms of bond breaking and bond formation the energy transfers in an exothermic reaction.

3 In a reaction the energy needed to break bonds is 300 kJ/mol and the energy transferred during bond formation is 200 kJ/mol. Is the reaction exothermic or endothermic? Give a reason for your answer.

4 What is activation energy?

6.2	
Co-ordinated	Modular
11.15	07 (10.19)

Reversible reactions

Some chemical changes are permanent whilst others can be easily reversed (see section 2.2). An example of a permanent chemical change is seen when paper is burned. The materials that make up the piece of paper combine with oxygen from the air in an exothermic reaction producing carbon dioxide and water vapour and leaving an ash. To reverse this process in the laboratory would be impossible. This is called an irreversible reaction.

Some reactions are fairly easy to reverse, for example the action of heat on solid ammonium chloride (NH_4Cl).

When ammonium chloride is heated as shown in Figure 6.3, the open end of the test tube remains cold whilst the solid itself becomes very hot and decomposes into two new products, ammonia (NH_3) and hydrogen chloride (HCl). These are both gases and they diffuse along the tube towards the cold end where they recombine to form solid ammonium chloride again. This is an example of a **reversible reaction**.

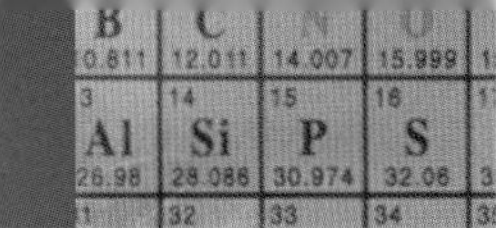

A reversible reaction must not be confused with similar physical changes that occur. For example, when paraffin wax is heated it first melts and then boils. If the paraffin wax vapour is cooled it condenses and then solidifies. This is quite different because when wax is heated it does not change chemically into a new substance. At all stages in the process the wax is present as wax molecules in either solid, liquid or gas state. With ammonium chloride, thermal decomposition (see section 2.3) of the solid into completely new compounds takes place and this is followed on cooling by the reverse chemical change taking place.

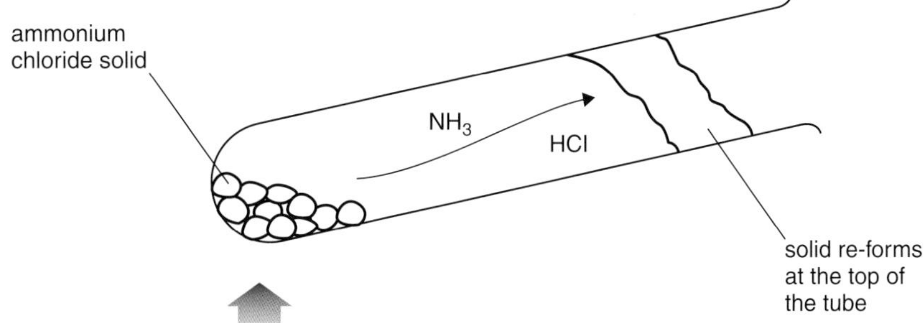

Figure 6.3
The action of heat on solid ammonium chloride

A reversible reaction is one that can take place in either direction depending on the conditions. Chemical reactions that are reversible have a special sign ($\rightleftharpoons$) in the equations that represents the change. If a reversible reaction is exothermic in one direction, it is endothermic in the other direction.

$$NH_4Cl(s) \underset{\textit{cool}}{\overset{\textit{heat}}{\rightleftharpoons}} NH_3(g) + HCl(g)$$

The laboratory test for water

Another example of a reversible reaction is the action of heat on hydrated (blue) copper sulphate crystals (see section 4.4). These have the formula $CuSO_4.5H_2O$ and when heated, the water of crystallisation separates and a white powder with the formula $CuSO_4$ is produced. This powder is called **anhydrous** copper sulphate. It is often used to detect the presence of water. When water is added to it, the colour changes from white back to the original blue. The reversible reaction can be summarised in the following equation:

$$\underset{\text{blue}}{CuSO_4.5H_2O} \rightleftharpoons \underset{\text{white}}{CuSO_4} + 5H_2O$$

Reversible reactions in a closed system

If the reversible reaction:

$$A + B \rightleftharpoons C + D$$

were to take place in a closed container so that none of the products could escape, after a while it would seem that nothing was happening. This would be because the reaction $A + B \rightarrow C + D$ was happening at exactly the same rate as the reaction
$C + D \rightarrow A + B$. The reversible reaction would now be described as being in a state of **equilibrium**.

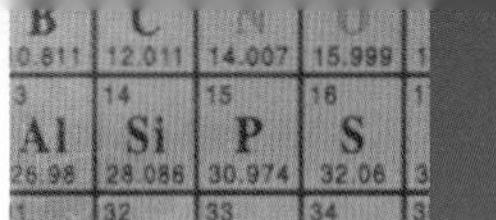

What can affect equilibrium?

1 Changing the temperature

- If the forward reaction is *endothermic* and the temperature is *increased*, the yield of products is *increased.*
- If the forward reaction is *endothermic* and the temperature is *decreased*, the yield of products is *decreased.*
- If the forward reaction is *exothermic* and the temperature is *increased*, the yield of products is *decreased.*
- If the forward reaction is *exothermic* and the temperature is *decreased*, the yield of products is *increased.*

2 Changing the pressure on reactions that involve reacting gases

- If there are more molecules on the left of the balanced symbol equation, then increasing the pressure will increase the yield of product.
- If there are more molecules on the right of the balanced symbol equation, then increasing the pressure will decrease the yield of product.
- If there are equal numbers of molecules on each side of the balanced symbol equation, then increasing the pressure will have no affect on the yield of product.

These changes in conditions and reaction rates are important in determining the optimum conditions in many industrial processes, including the Haber process (see section 3.2).

Summary

- Some reactions can proceed in either direction depending on the conditions. These are **reversible reactions**.
- When a reversible reaction occurs in a closed system, an **equilibrium** is reached when the reactions occur at exactly the same rate in each direction.

Topic questions

1 What is the difference between a permanent chemical change and a reversible reaction?

2 Describe what happens when ammonium chloride is placed at the bottom of a long glass tube and heated?

3 a) What is the chemical test for water?
b) How can this reaction be made to reverse?

6.3	
Co-ordinated	Modular
11.13	07 (10.17)

Rates of reaction

The rate of a chemical reaction is the speed at which it takes place. It can vary from a fraction of a second to several weeks, months or even years and it is not always possible to predict how fast a given reaction will take place.

- Some reactions take place very quickly e.g. an explosion.
- Some take place quite quickly e.g. the action of water on potassium metal.
- Some take place quite slowly e.g. the rusting of iron.
- Some take place very slowly e.g. the fermentation of grapes into wine.

The only real way to find out how fast a chemical reaction takes place is to carry it out.

Do not confuse how *fast* a reaction takes place with how *much* reaction takes place. The amount of reaction taking place depends on the amounts of chemicals used. If marble chippings (calcium carbonate) are added to dilute hydrochloric acid, the gas carbon dioxide is given off. The actual volume of gas produced depends on how much marble and how much acid was used.

- If a lot of marble and acid are used, a lot of carbon dioxide is produced.
- If a small piece of marble and a lot of acid are used, the marble is likely to be used up. Some of the acid will remain unreacted. The amount of gas produced here depends on the amount of marble used.
- If a huge lump of marble and just a small amount of acid are used, the acid will be used up and some of the marble will remain. The amount of gas produced this time depends on the amount of acid used.

The amount of reaction is determined by the material that is present in *least* quantity.

Factors affecting the rate of a reaction

The rate at which a reaction takes place depends on a number of factors. Several things can be done to either speed up or slow down a chemical reaction. These are:

- changing the *surface area* of any solids involved in the reaction
- changing the *concentration* of any reactants that are in solution
- changing the *pressure* in reactions where gases are involved
- changing the *temperature* of the reaction mixture
- adding a *catalyst*.

For each of these factors it is important to realise exactly how and why the reaction rate changes.

Factors that can have their values changed are known as variables. If the value of one variable (the independent variable) is changed, then the value of a second variable (the dependent variable) will also change. To find out how changing independent variables affects a dependent variable it is essential that we change only one variable at a time.

In rates of reaction experiments, if more than one independent variable is altered at a time it is impossible to decide which one is causing the change in the reaction rate. The idea of changing one independent variable at a time is used in each of the experiments described in this chapter.

Changing the surface area

This factor only affects reactions where a solid is involved. Figure 6.5 shows what happens on the surface of a piece of marble when it is dropped into dilute hydrochloric acid.

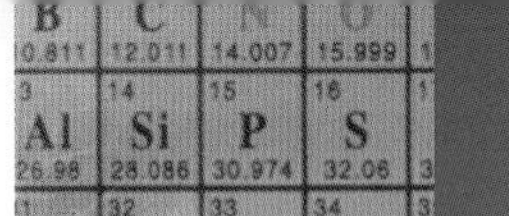

The particles that really cause the reaction are the hydrogen ions ($H^+(aq)$) that come from the acid. These ions are able to move freely through the solution and will eventually bump into the surface of the piece of marble, reacting with it. The marble in the centre of the lump has to wait until all the marble around it has been reacted before it has a chance to react with the H^+ ions.

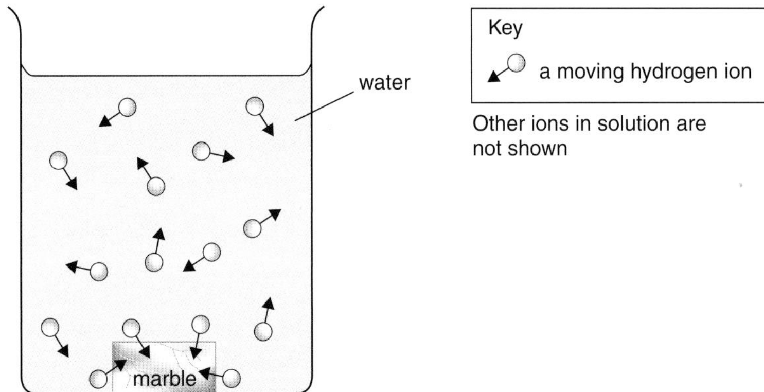

Figure 6.4
Hydrogen ions colliding with a lump of marble

Each time a hydrogen ion collides with the marble it reacts and is neutralised. The number of hydrogen ions steadily decreases reducing the concentration of the acid. As this happens the rate of the reaction slowly decreases. If the marble is in excess, eventually all the hydrogen ions are used up and the reaction stops altogether. If the experiment is repeated with the same amount of marble broken into several smaller pieces, more of the marble that was in the centre of the large lump will become exposed to the acid and will be able to react immediately and not have to wait for the marble around it to be dissolved first (see Figure 6.5).

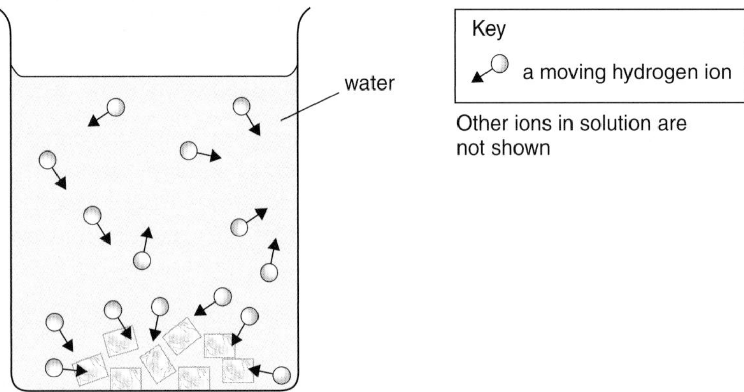

Figure 6.5
Hydrogen ions colliding with smaller pieces of marble

There are several applications where changing the particle size can alter the rate of a reaction.

- The burning of coal in power stations can be speeded up by converting it into a very fine powder. It can then be blown through a pipe where it burns as if it were a gas. Using a powder speeds up the combustion process.
- Large pieces of graphite are used as anodes in the extraction of aluminium by electrolysis (see section 4.2). This slows down the rate at which the graphite burns in the atmosphere of oxygen that is generated around the anodes during the process. Using large lumps slows down the combustion process.

For a given amount of material, the smaller the size of the pieces, the larger the surface area and the faster reactions will take place.

The effect of particle size/surface area on the rate of a chemical reaction can be demonstrated using the reaction between marble and dilute hydrochloric acid (see Figure 6.6).

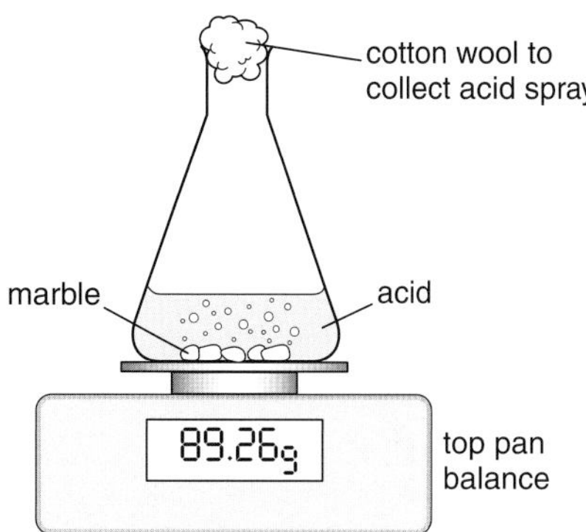

Figure 6.6
The reaction between acid and marble

A known mass of large pieces of marble is dropped into a known volume of dilute hydrochloric acid in a conical flask. A plug of cotton wool is placed in the neck of the flask to trap any acid spray but allow any gas to escape. As the gas escapes, the mass of the apparatus decreases. The decrease in mass is recorded each minute until there is no further change, i.e. until the reaction has stopped.

The experiment is then repeated using the same volume of acid of the same concentration and at the same temperature. The same mass of marble pieces is used but the size of the pieces is changed. In one repeat, medium-sized pieces are used; in the next, small pieces are used. (Note that the only variable changed here is the size of the particles being used.) Figure 6.7 shows a typical set of results obtained and Figure 6.8 shows these results in the form of three graphs.

Time (mins)	Total mass loss (g)		
	Large pieces	Medium pieces	Small pieces
0	0.00	0.00	0.00
1	1.00	1.48	2.96
2	1.80	2.52	3.68
3	2.52	3.20	3.85
4	3.00	3.58	3.93
5	3.29	3.73	3.98
6	3.50	3.85	3.99
7	3.66	3.92	4.00
8	3.76	3.96	4.00
9	3.82	3.99	4.00
10	3.88	4.00	4.00
11	3.93	4.00	4.00
12	3.97	4.00	4.00
13	3.99	4.00	4.00
14	4.00	4.00	4.00
15	4.00	4.00	4.00

Figure 6.7
Table of results

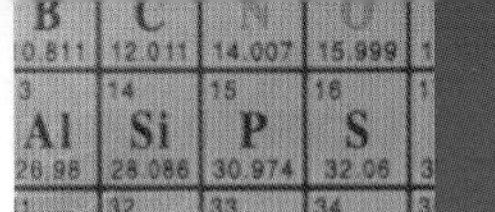

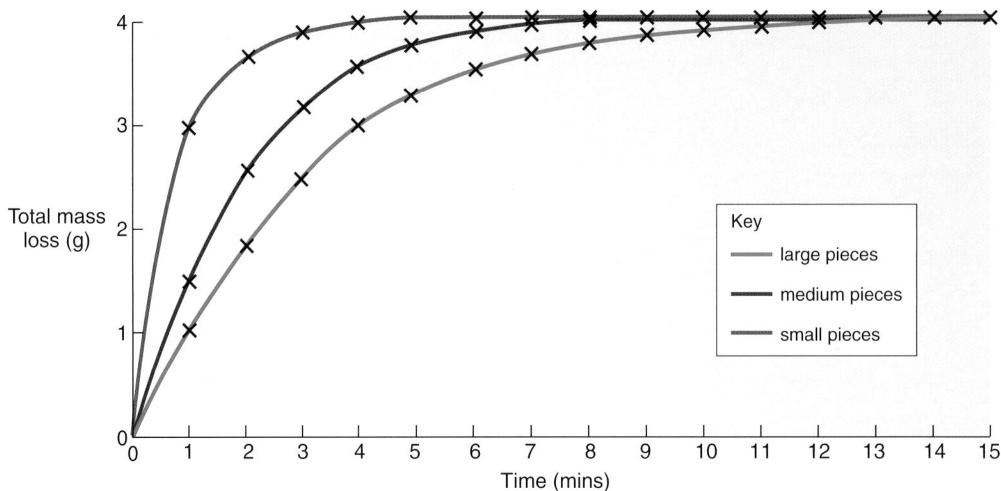

Figure 6.8
The effect of chip size on reaction rate

Notice how in Figure 6.8 each graph eventually reaches the same height but at a different time. The same overall mass loss is obtained because equal amounts of reactants have been used in all three experiments. The slope of the initial part of each graph is an indication of the rate of the reaction. The steeper the slope, the faster the reaction. So the smaller the marble pieces, the faster the reaction. This is because smaller pieces have a larger surface area.

Did you know?

Smoking is not allowed in flour mills partly because of the danger of an explosion caused by the extremely fine particles of flammable material (flour) floating in the air. If these particles caught fire they would burn like a gas and the flame could spread very quickly! Workers wear rubber-soled shoes and sometimes the mills even have rubber floors to prevent the risk of a spark.

Changing the concentration of reactants in solution

The rate of a reaction can also be altered by using solutions of different concentrations. Figure 6.9 shows the effect of doubling the concentration of the acid. There are now twice as many particles in the same volume of solution. This means there is twice the chance of these acid particles colliding with the marble and causing a chemical reaction to occur.

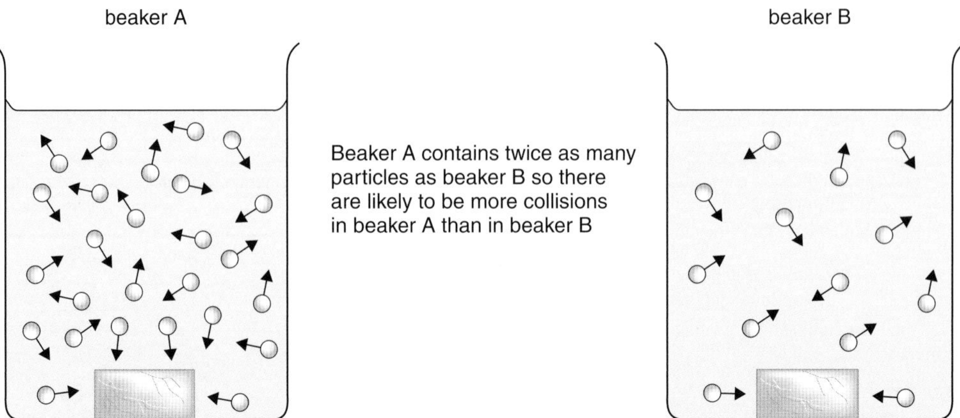

Figure 6.9
Changing the concentration of a substance in solution

In very dilute solutions of soluble substances, the particles are a long way apart and will have to travel greater distances before they can collide and react. This is going to take time, so reactions involving dilute solutions are slow.

The effect of changing the concentration of a solution on the reaction rate can be followed by measuring the volume of gas given off over a period of time. Equal volumes of solutions of hydrochloric acid of different concentrations are added to the same mass of small marble chips.

A typical set of apparatus for doing this is shown in Figure 6.10. (The graduated measuring cylinder shown could be replaced by a gas syringe or burette.)

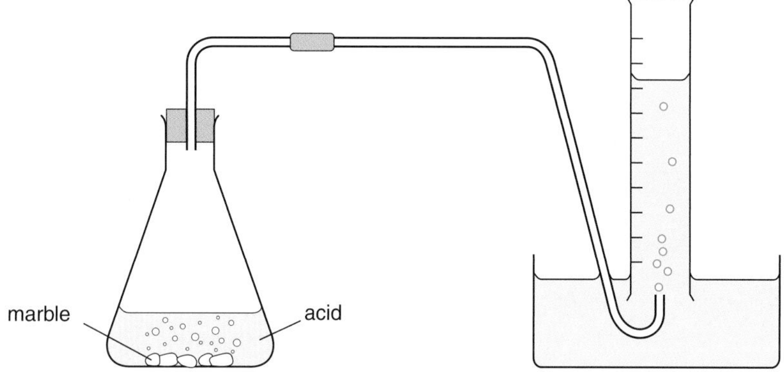

Figure 6.10
Collecting and measuring the volume of gas evolved in a reaction

Readings of the volume of gas collected after each minute are made until the reaction is complete. The reaction stops in each case when the marble chip has completely dissolved.

Typical readings using solutions of three different concentrations are shown in Figure 6.11.

Figure 6.11
Table of results

Time (mins)	Concentration of acid used		
	Low concentration	Medium concentration	High concentration
0	0.0	0.0	0.0
1	3.5	5.5	10.0
2	7.0	10.5	16.0
3	10.0	14.5	20.5
4	13.0	18.0	23.0
5	15.5	20.5	24.7
6	18.0	22.5	25.7
7	20.1	24.0	26.3
8	21.9	25.0	26.5
9	23.5	25.5	26.5
10	25.0	26.0	26.5
11	26.0	26.5	26.5
12	26.5	26.5	26.5

The graphs of these results (shown in Figure 6.12) are similar in shape to those for changing the particle size. Once again the steeper the slope of the graph, the faster the reaction is taking place. The graphs show that as the concentration of the acid increases, the rate of reaction also increases.

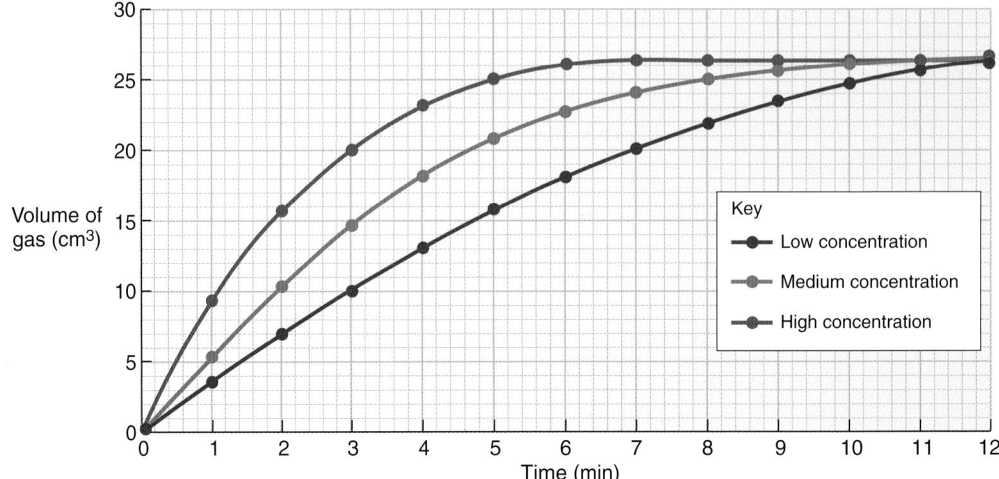

Figure 6.12
The effect of concentration on reaction rate

Another useful reaction to use to study the effect of changing the concentration of a solution on the reaction rate is that between sodium thiosulphate solution ($Na_2S_2O_3(aq)$) and dilute hydrochloric acid ($HCl(aq)$).

The equation for the reaction is:

$$Na_2S_2O_3(aq) + 2HCl(aq) \rightarrow 2NaCl(aq) + H_2O(l) + SO_2(g) + S(s)$$

During the reaction solid sulphur is produced but it appears very slowly and at first there appears to be no reaction. After a few moments, however, the mixture slowly starts to turn cloudy and if the reaction vessel is placed over a pencilled cross drawn on a piece of paper (see Figure 6.13) the cloudiness gradually obscures the cross until it can no longer be seen. The time taken for this to happen can be used as the basis for following the reaction.

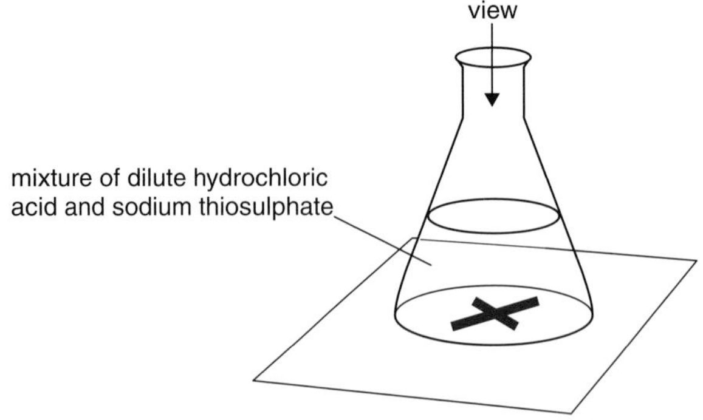

Figure 6.13
Sodium thiosulphate and hydrochloric acid

In the experiment, as the sodium thiosulphate is diluted, the reaction particles become more spread out so they will have to travel further before they can react. Figure 6.14 lists some typical results obtained. The longer a reaction takes, the slower it is. The rate of reaction is 'inversely proportional' to the time taken. This means that:

$$\text{rate of reaction is proportional to } \frac{1}{\text{time}}$$

Figure 6.14

Experiment number	Volume of sodium thiosulphate used (cm^3) / Concentration	Volume of water used (cm^3)	Volume of hydrochloric acid used (cm^3)	Time (sec)	1/time (sec^{-1}) / Rate
1	50	0	5	38	0.0263
2	40	10	5	47	0.0213
3	30	20	5	62	0.0161
4	25	25	5	74	0.0135
5	20	35	5	95	0.0105
6	10	40	5	182	0.0055

The last column in the table shows the value of 1/time which is a measure of the rate of the reaction. Figures 6.15 and 6.16 show the graphs produced when the 'concentration' of the thiosulphate solution is plotted against 'time' and also against '1/time'.

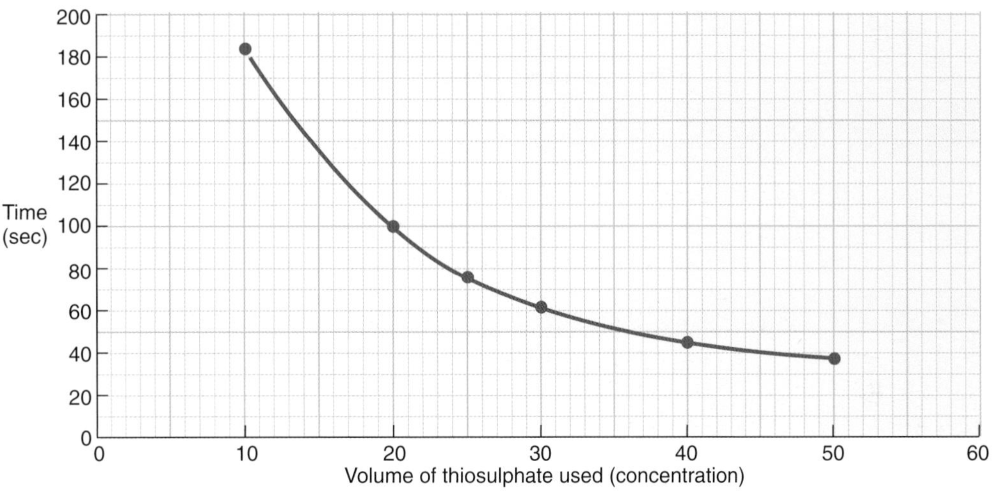

Figure 6.15 *The effect of concentration on the time of a reaction*

Note that in Figure 6.16, as the concentration of the thiosulphate solution increases, the rate of reaction increases.

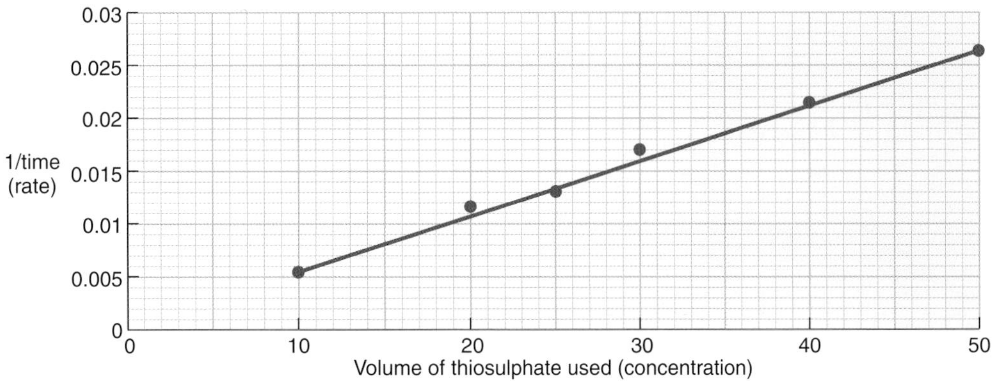

Figure 12.16 *The effect of concentration on rate of reaction*

Changing the temperature

The kinetic theory says that when the temperature of a substance is increased, the particles in a solid vibrate more and the particles in liquids and gases move faster.

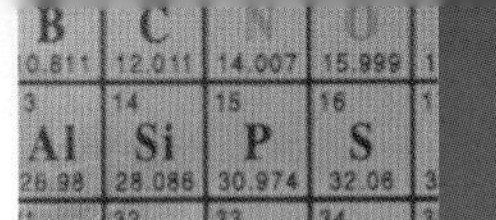

Chemistry in action

When a reaction mixture is heated up, each particle in it acquires more energy. The particles move around faster so that collisions between them are likely to happen more often. This increases the reaction rate for two reasons:

- the number of collisions taking place will increase
- the number of collisions where the energy of the colliding particles exceeds the energy of activation for the reaction will increase.

The effect of changing the temperature on the reaction rate can be shown using the reaction between sodium thiosulphate solution and dilute hydrochloric acid. The equipment and method of following the reaction used would be similar to that described earlier in this chapter (see Figure 6.13). A number of reactions would be carried out in which the concentrations and volumes of sodium thiosulphate solution and dilute hydrochloric would be kept the same but the temperature of each reaction would be different. The time taken for the cross to disappear when viewed from above through the solution would be recorded. Some typical results are shown in Figure 6.17.

Figure 6.17

Temperature at start of reaction (°C)	Time for cross to disappear (sec)
47	33
42	38
38	49
35	66
30	100
21	181

These results are shown in the form of a graph in Figure 6.18. Note how the time for the reaction to take place decreases as the temperature increases.

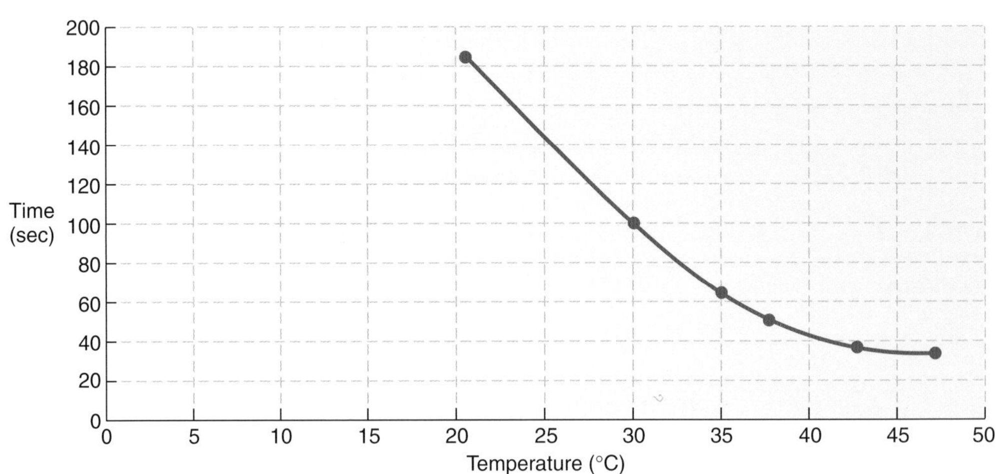

Figure 6.18
The effect of temperature on reaction time

Adding a catalyst

A **catalyst** is a substance that can alter the rate of a chemical reaction but is not used up and remains chemically unchanged at the end of the process. Catalysts are used extensively in industry and most modern cars now incorporate them in catalytic converters to convert harmful exhaust gases to comparatively harmless ones before they are released into the atmosphere.

Some reactions are very slow, for example the gradual decomposition of hydrogen peroxide (H_2O_2) into water and oxygen.

$$2H_2O_2(aq) \rightarrow 2H_2O(l) + O_2(g)$$

This reaction can be speeded up by adding manganese(IV) oxide ($MnO_2(s)$) as a catalyst. When the two substances are mixed there is a much more vigorous reaction and oxygen is readily given off. Tiny amounts of catalyst can bring about very large changes in reaction rates. Not all reactions can be speeded up in this way. As the amount of catalyst used increases, the rate of reaction also increases until a point is reached where the reaction will go no faster.

If increasing amounts of catalyst are added to hydrogen peroxide solution, the time to produce 50 cm^3 of oxygen decreases as shown in Figure 6.19. To test a gas to see if it is oxygen, put a glowing spill into the gas. If it is oxygen, the spill will relight.

Figure 6.19
The effect of amount of catalyst on reaction rate

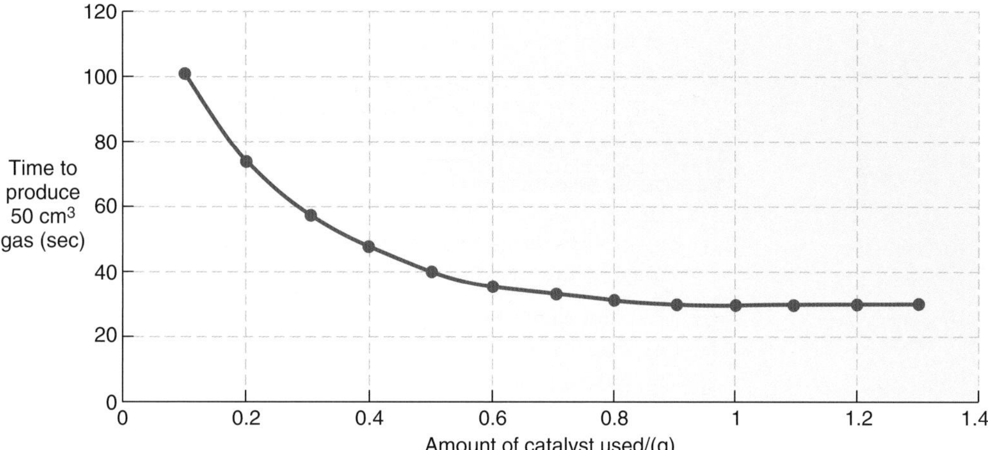

Note that the addition of more than 0.9 g of catalyst does not produce any further increase in the reaction rate.

Catalysts work by providing an easier route from reactants to products and this is done by lowering the energy barrier – i.e. reducing the energy of activation for the reaction.

Figure 6.20
An energy level diagram for a reaction involving a catalyst

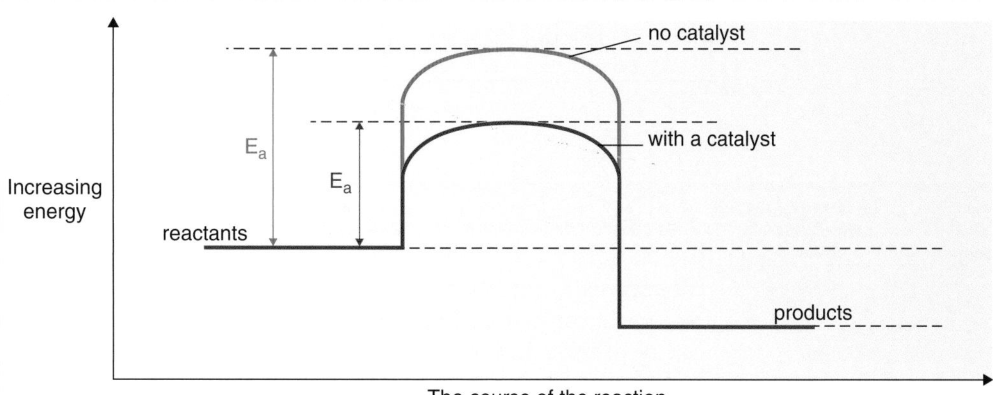

Figure 6.20 shows the energy level diagram for a reaction both with and without a catalyst. The catalyst has provided a lower energy barrier and therefore it is easier for the particles to change into products.

Did you know?

Whilst catalysts are not used up during chemical reactions they can be easily 'poisoned'. This happens if they come in contact with substances that block the active sites on their surface. If, for example, leaded petrol is used in a car that has a catalytic converter, the surface of the catalyst can become covered in lead compounds and this will prevent the gases getting absorbed on to the surface and therefore the catalytic activity is destroyed.

Summary

- The rate of a chemical reaction can be altered by:
  - changing the surface area of any solids involved
  - changing the concentration of any solutions involved
  - changing the temperature at which it is carried out
  - changing the pressure, if gases are involved
  - the use of a **catalyst.**
- A catalyst is a substance that can alter the rate of a reaction but is not used up in the process.

Topic questions

1 When magnesium ribbon is added to dilute sulphuric acid it starts to dissolve and give off hydrogen gas. Suggest three ways in which the rate at which the magnesium dissolves could be increased.

2 Explain each of the following:
a) Sugar dissolves faster in hot water than in cold water.
b) When producing carbon dioxide gas by adding marble to an acid, the reaction is quickest when the marble is in powdered form.

3 Why does increasing each of these variables increase the rate of a reaction?
a) surface area
b) concentration
c) temperature
d) pressure

4 When the temperature increases the rate of reaction increases. What happens to the amount of products produced? Give a reason for your answer.

5 a) What does adding a catalyst do to the rate of a reaction?
b) How does the addition of a catalyst affect the activation energy level for the reaction?

6.4 Reactions involving enzymes

Co-ordinated	Modular
11.14	07 (10.18)

Enzymes are complicated molecules made up of proteins. They are often described as biological catalysts. Most catalysts contain metals – often transition metals from the centre of the periodic table. Enzymes usually do not contain metals.

Enzymes act as catalysts in a different way to metal and metal compound catalysts. Each enzyme has a particular shape, so each enzyme can only combine with molecules that have a matching shape. The enzyme and its matching molecule lock together like the pieces of a jigsaw. This is called the 'lock and key' model of enzyme action – because only the matching molecule (the key) will fit the shape of the enzyme (the lock).

The fact that enzymes will only react with molecules of a particular shape explains why each enzyme is specific to a particular reaction.

Once the reactant molecule (also called the **substrate**) combines with the enzyme, it can change to a product molecule. This product will fall away from the enzyme. Therefore enzymes can be used again and again without being used up. This can be seen in Figure 6.21.

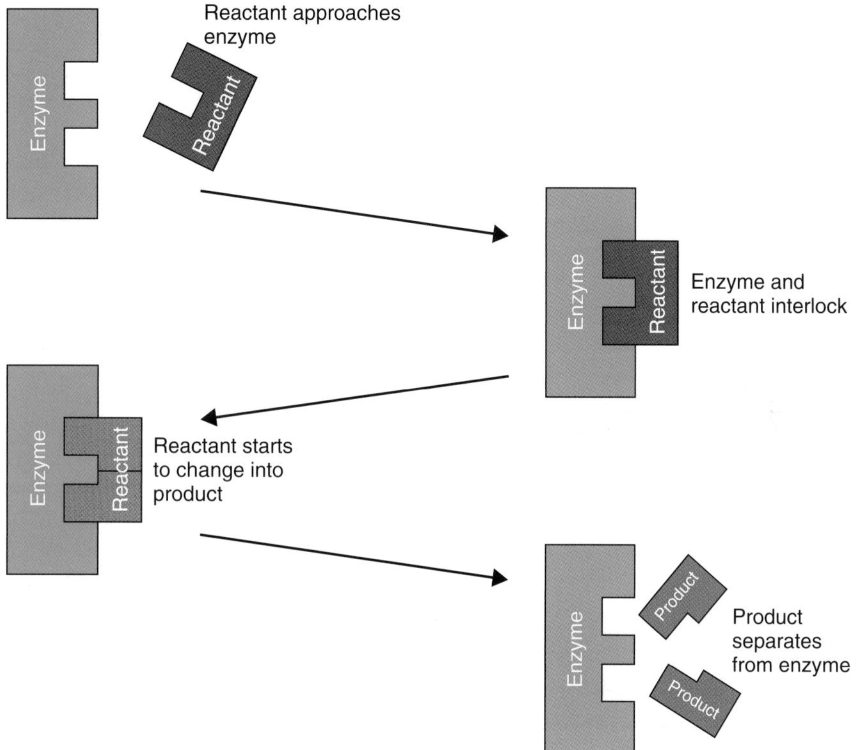

Figure 6.21
How enzymes catalyse reactions

Examples of enzymes can be found in the digestive system where they break down large molecules of starch, protein and fats into smaller molecules which can be absorbed by the bloodstream. Amylase is the enzyme, which breaks down starch into glucose, proteases break down proteins into amino acids and lipases break down fats to fatty acids. In these examples, the enzymes (amylase, protease, and lipase) are the 'locks' and glucose, protein and fats are the 'keys'.

The chemical composition of enzymes, and hence their shape and catalytic activity, can be easily destroyed by heating. This is called **denaturing**. Enzyme activity is poor at low and high temperatures but is quite vigorous at temperatures around 25 to 35°C. Enzymes are therefore only able to catalyse reactions below about 50°C.

Different enzymes work best at different pH values.

Reactions involving enzymes

Three important processes that involve enzymes are making bread, brewing beer and wine and making yoghurt.

Making bread

The ingredients for making bread include flour, sugar, salt and yeast. Yeast is a naturally-occurring material that lives off decaying plant or animal material. Yeast does not contain chlorophyll so cannot make its own food but it does contain enzymes.

If yeast is mixed with the other bread-making ingredients and left in a warm place, the mixture (the dough) begins to rise because carbon dioxide is given off and the warmth makes the gas expand. This process is called **fermentation**. The reaction that takes place is:

$$\underset{\text{glucose}}{C_6H_{12}O_6} \rightarrow \underset{\text{ethanol}}{2C_2H_5OH} + \underset{\text{carbon dioxide}}{2CO_2}$$

The risen dough is then baked in an oven where the high temperature destroys the enzymes in the yeast, makes the carbon dioxide gas bubbles expand even further and causes the bread to rise.

Making beer and wine

Another common process which uses enzymes is the conversion of sugar into ethanol (alcohol). This process is also a fermentation reaction. It requires warm conditions – high temperatures destroy the enzymes present. The reaction is the same as the one for making bread.

$$\underset{\text{glucose}}{C_6H_{12}O_6} \rightarrow \underset{\text{ethanol}}{2C_2H_5OH} + \underset{\text{carbon dioxide}}{2CO_2}$$

Many of the starting materials for making wine or beer contain starch. Enzymes in the yeast are responsible for catalysing the breakdown of the complex organic starch molecules into simpler sugars such as sucrose and glucose. These are then converted into ethanol (alcohol) and carbon dioxide. As the fermentation process proceeds, bubbles of carbon dioxide gas are evolved and if a still wine is required, all of this gas is allowed to escape. Where a fizzy wine or beer is required, a secondary fermentation process is allowed to take place. This time the gas is not allowed to escape until the bottle is opened.

Making yoghurt

This process involves allowing the enzymes in the bacteria in unpasteurised milk to convert the natural sugar present in the milk – lactose – into the acid lactic acid. This makes the milk taste sour but it also helps to preserve it. If pasteurised milk is used – and this is more common nowadays – then yoghurt makers have to add their own strains of bacteria in order to make the process work. The process is carried out under warm conditions because high temperatures destroy the enzymes.

Uses of enzymes in the home and industry

Figure 6.22
Some uses of enzymes

Use	Enzyme	Action	Comments
Biological washing powders	lipases proteases	remove fat-based stains remove protein-based stains	will work at low temperatures
Baby foods	proteases	pre-digest protein	the baby's digestive system has less work to do
Sweeteners	isomerase	converts glucose into fructose	fructose is very much sweeter than glucose, so less needs to be used
Production of sugar syrup	carbohydrases	converts starch syrup obtained from plants into maltose and glucose both of these sugars are sweet	different amounts of enzyme produce either a syrup containing more maltose than glucose – used in the brewing industry – or a syrup containing more glucose than maltose. Further enzyme action on the latter turns the glucose into fructose and produces a syrup used in foods and drinks

Enzyme technology

Figure 6.23
Fermentation tanks

The enzymes used in industry are usually extracted from microbes (bacteria, moulds and fungi). It is now possible for scientists to 'programme' bacteria to make certain enzyme drugs. Human insulin can now be manufactured in this way.

The use of enzymes in industry has the benefit of allowing chemical reactions that would otherwise require the use of high temperatures and/or high pressures to take place at normal temperatures and pressures. Their use also means that the chemical reactions are likely to be less costly and that the reactions can be allowed to take place as a continuous process.

However, a successful product will only be produced if certain conditions can be guaranteed. These include:

- That the bacteria, which are grown in large quantities from which the enzymes are to be extracted, are all identical. This is necessary to ensure that the extracted enzymes are identical.
- That the enzymes must be identical to the strain required. This is necessary to ensure that the correct chemical reaction is controlled and that unwanted products are not produced.
- That there is close monitoring of the reaction conditions to ensure that a safe and uncontaminated product results.
- That all modified bacteria are contained safely until it is proven that they are no risk to any other organisms.

Getting the best from the enzymes

Any enzyme used in industry must be pure and not contaminated in any way that will affect its action as a biological catalyst, and must be able to be made in large quantities.

The required enzyme is extracted and purified and allowed to reproduce in sterile conditions in a fermenter. (A large tank where pH, temperature and food source can be monitored and regulated so that very large quantities of the identical enzyme can be produced.)

Whilst the enzyme is carrying out its intended function the main problem is to make sure that it stays working for as long as possible by controlling and monitoring the immediate working environment. This is called stabilisation.

In order for the reactants involved in the process to come into contact with the enzyme in a continuous process the enzyme needs to be immobilised. This can be achieved by trapping the enzyme in an inert solid support or a carrier made of alginate beads. (Alginate is an extract of seaweed, similar to agar.)

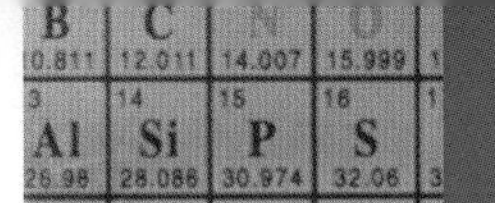

Summary

- **Enzymes** are complex organic molecules that can catalyse specific reactions. They are biological catalysts that can be destroyed by heating.
- Enzymes are used in the manufacture of alcohol, bread and yoghurt.
- Enzymes are used in industry to control reactions that would otherwise require high temperatures and/or pressures.

Topic questions

1 a) When a piece of fresh liver is added to hydrogen peroxide solution (H_2O_2) there is a vigorous reaction and oxygen gas is given off. If in a second similar experiment a piece of liver that has been boiled in water for 5 minutes is used, no gas is produced. Explain this reaction.
b) Explain why food keeps for longer periods of time if stored in a refrigerator.

2 What are enzymes?

3 Use the 'lock and key' model to explain how enzymes work.

Examination questions

1 Some types of filler go hard after a catalyst is added from a tube. A manufacturer tested this reaction to see what effect the amount of catalyst had on the time for the filler to harden. The results are shown in the table.

Volume of catalyst added to filler (cm^3)	Time for the filler to harden (minutes)
1	30
2	15
3	10
4	7
6	4

a) Draw a graph of these results. Plot 'time for the filler to harden', in minutes, on the vertical axis and 'volume of catalyst added to filler', in cm^3, on the horizontal axis. *(3 marks)*

b) Use your graph to suggest the time taken for the filler to harden using 5 cm^3 of catalyst. *(1 mark)*

c) What is the effect of the catalyst on the rate of this reaction? *(1 mark)*

2 Hydrogen peroxide, H_2O_2, is often used as a bleach. It decomposes forming water and oxygen.

a) Write the balanced chemical equation for the decomposition of hydrogen peroxide. *(3 marks)*

b) The rate of decomposition of hydrogen peroxide at room temperature is very slow. Manganese oxide is a catalyst which can be used to speed up the decomposition. Complete the sentence.
A catalyst is a substance which speeds up a chemical reaction. At the end of the reaction, the catalyst is ________ . *(1 mark)*

c) Two experiments were carried out to test if the amount of manganese oxide, MnO_2, affected the rate at which the hydrogen peroxide decomposed.

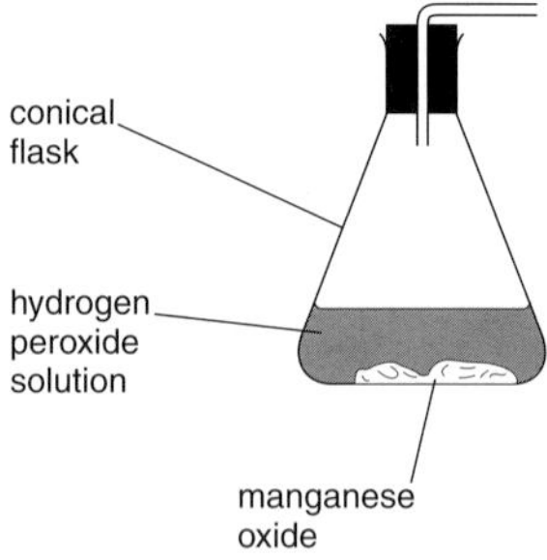

i) Complete the diagram to show how you could measure the volume of oxygen formed during the decomposition. *(2 marks)*

ii) The results are shown in the table below. Draw a graph of these results. The graph for 0.25 g MnO_2 has been drawn for you. *(3 marks)*

iii) Explain why the slopes of the graphs become less steep during the reaction. *(2 marks)*

iv) The same volume and concentration of hydrogen peroxide solution was used for both experiments. What *two* other factors must be kept the same to make it a fair test? *(2 marks)*

Time in minutes	0	0.5	1	1.5	2	2.5	3	3.5
Volume of gas in cm^3 using 0.25 g MnO_2	0	29	55	77	98	116	132	144
Volume of gas in cm^3 using 2.5 g MnO_2	0	45	84	118	145	162	174	182

3 a) i) A student added a few drops of Universal Indicator to some sodium hydroxide solution. The indicator turned purple. What was the pH of the sodium hydroxide solution? *(1 mark)*

ii) The student added an acid until the solution was neutral. What is the colour of Universal Indicator when the solution is neutral? *(1 mark)*

b) i) Which acid should the student add to sodium hydroxide solution to make sodium sulphate? *(1 mark)*

ii) Write the formula of sodium sulphate. *(1 mark)*

c) The student noticed that the solution in the beaker got warm when the acid reacted with the alkali. The energy diagram below represents this reaction.

i) In terms of **energy**, what type of reaction is this? *(1 mark)*

ii) Use the energy diagram to calculate a value for the amount of energy released during this reaction. *(1 mark)*

iii) Explain, in terms of bond breaking and bond forming, why energy is released during this reaction. *(3 marks)*

iv) The reaction takes place very quickly, without the help of a catalyst. What does this suggest about the activation energy for this reaction? *(1 mark)*

Energy (kJ)
(1 small division = 10 kJ)

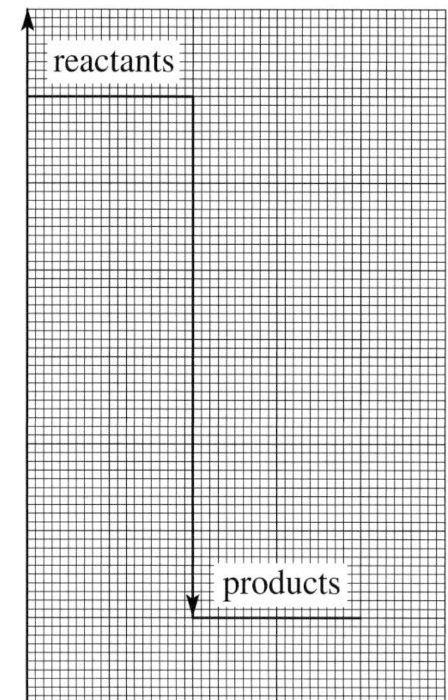

4 This item appeared in the Wolverhampton *Express and Star* on October 31st, 1997. Read the passage and answer the questions that follow.

Fumes scare at factory

Workers were forced to flee a factory after a chemical alert. The building was evacuated when a toxic gas filled the factory.
It happened when nitric acid spilled on to the floor and mixed with magnesium metal powder.

a) The equation which represents the reaction between magnesium and nitric acid is:

$$Mg(s) + 4HNO_3(aq) \rightarrow Mg(NO_3)_2(aq) + 2H_2O(l) + 2NO_2(g)$$

Give the formula of the toxic gas that was produced. *(1 mark)*

b) What does toxic mean? *(1 mark)*

c) The reaction of nitric acid with magnesium metal powder is more dangerous than if the acid had fallen on to the same mass of magnesium bars. Explain why. *(1 mark)*

d) i) Water was sprayed on to the magnesium and nitric acid to slow down the reaction. Explain, in terms of particles, why the reaction would slow down. *(2 marks)*

ii) Explain why it is better to add alkali, rather than just add water to the spillage. *(1 mark)*

Chapter 7

Organic chemistry

Key terms addition polymerisation • addition reactions • **alcohol** • alkanes • alkenes • **anaerobic** • anion • **biological catalyst** • boiling point • bond energy • branched chain • carboxylic acids • cation • **complete combustion** • covalent • cross-linking bonds • double bonds • **enzyme** • ester • **fermentation** • functional group • general equation • general formula • **haemoglobin** • homologous series • hydrogenated • **incomplete combustion** • ionic • isomers • melting point • molecular formula • monomers • **organic compounds** • **oxidise** • plastics • polymer • relative molecular mass • saturated hydrocarbons • steroid • straight chain • structural formula • thermosetting plastics • thermosoftening plastics • unsaturated hydrocarbons • **yeast**

7.1 The meaning of 'organic compounds'

Co-ordinated	Modular
10.4	21 (14.6)

Most **organic compounds** come from living organisms but some can be made from inorganic materials. The compounds in crude oil, natural gas and coal are organic because they are the fossilised remains of living organisms. Living materials like wood also contain organic compounds.

All organic compounds contain the element carbon. Most of them also contain hydrogen and some contain other elements like oxygen, nitrogen, sulphur or the halogens.

Did you know?

Urea is a compound found in animal urine so it is an organic compound. It is used as a fertiliser and in the manufacture of certain types of polymers. But urea can also be manufactured commercially from the inorganic materials carbon dioxide and ammonia.

Figure 7.1
These substances – wood, coal and methylated spirit – are all organic

7.2	
Co-ordinated	Modular
10.4	21 (14.6)

Burning organic compounds

When organic compounds are burned in air, the hydrogen is always **oxidised** to water (see section 4.4).

If the air supply is plentiful, **complete combustion** takes place and the carbon is oxidised to carbon dioxide. But if the air supply is limited, **incomplete combustion** takes place and carbon monoxide and/or the element carbon are produced.

Carbon monoxide is highly poisonous because it combines with the **haemoglobin** in the blood and stops it from binding with oxygen. The body becomes starved of oxygen and this can lead to unconsciousness and death.

When incomplete combustion of an organic compound produces carbon, the small carbon particles get heated in the flame and glow. This makes the flame bright yellow. Once the particles are out of the flame they cool quickly and become black smoke. So a yellow, smoky flame is a sign of incomplete combustion.

With incomplete combustion, only some of the chemical energy from the organic compound is transferred to heat energy. Incomplete combustion is, therefore, inefficient compared with complete combustion.

When the air hole of a Bunsen burner is closed, there is not enough oxygen for the complete combustion of the gas. The flame is yellow and smoky and takes a long time to heat a beaker of water. Once the air hole is opened, air is sucked in the hole and complete combustion occurs. The resulting blue flame is much better at heating.

Figure 7.2
The yellow Bunsen flame shows incomplete combustion of the gas. The blue Bunsen flame shows complete combustion

The more carbon atoms a molecule has, the more oxygen is needed to oxidise it fully (see Figure 7.3).

Compound	% carbon	Flame colour	Is combustion complete?	Is combustion efficient?
ethanol	52	blue	yes	yes
methane (natural gas)	75	blue or yellow	sometimes	varies
octane (in petrol)	84	yellow, smoky	no	no

Figure 7.3
The effect of carbon content on the colour of the flame

The complete combustion of compounds with large organic molecules requires a lot of air. Specialised furnaces and boilers are needed to burn these compounds efficiently. In most cases this is not an economic situation. But in large factories and oil-powered ships that require a lot of energy, the process is economically sensible because the high cost of the boilers is offset by the much lower cost of the fuel.

In the home the commonly used fuels are coal, oil and natural gas. The table below gives some data on these fuels.

Fossil fuel	Approx. energy output (kJ/g)	Approx. mass of CO_2 per 1000 kJ of energy (grams)	Main pollutants produced	Relative cost of fuel for domestic heating Natural gas = 1
coal – lignite coal – anthracite	17 35	100	smoke oxides of sulphur oxides of nitrogen	1.26
oil – petrol (gasoline) oil – paraffin (kerosene)	50 45	80	oxides of nitrogen	1.31
natural gas	90	60	oxides of nitrogen	1.00

Natural gas has several advantages over the other fuels. It produces more energy for the same mass of fuel and it produces less of the greenhouse gas carbon dioxide to get that energy. It is also considerably cheaper than the other fuels. Oil, though not as good as natural gas, is less polluting than coal. Oil does not have the high sulphur content of coal and it does not produce smoke. It also produces less carbon dioxide than coal to get the same amount of energy.

Equations for the combustion of methane

For the complete combustion of methane, the equation is:

$$CH_4 + 2O_2 \rightarrow CO_2 + 2H_2O$$

If combustion is incomplete, producing carbon monoxide, the equation is:

$$CH_4 + 1\tfrac{1}{2}O_2 \rightarrow CO + 2H_2O$$

The second equation needs less oxygen so it is the reaction that will occur when the oxygen supply is limited.

Organic compounds which contain nitrogen and chlorine, which are present in many polymers (plastics), burn to produce the toxic gases hydrogen cyanide (HCN) and hydrogen chloride (HCl), respectively. More of these gases is made if the air supply is limited.

Topic questions

1 What substances are formed when an organic compound burns completely in air?

2 Which of the following compounds would take most air for complete combustion?
a) C_2H_6
b) C_3H_6
c) C_5H_{12}
d) C_7H_{14}

3 Explain why the burning of some plastics produces toxic gases.

Summary

- Compounds derived from living organisms are **organic compounds**.
- All organic compounds contain carbon. Most of them also contain hydrogen.
- When organic compounds are burned, the hydrogen is always **oxidised** to water.
- If **complete combustion** takes place, carbon is oxidised to carbon dioxide.
- If **incomplete combustion** takes place, carbon is oxidised to carbon monoxide.
- Some organic compounds contain nitrogen or chlorine. When they burn, toxic gases are produced.

7.3	
Co-ordinated	Modular
10.4	21 (14.7)

Homologous series

An **homologous series** is a family of compounds with similar structures and chemical properties. All members of an homologous series have the same **general formula**. The alkanes, alkenes, alcohols and carboxylic acids are examples of homologous series.

The alkanes

The general formula of the **alkanes** is C_nH_{2n+2}. All alkanes are flammable but otherwise are fairly unreactive (see also section 4.4).

Figure 7.4 shows the **structural formulae** of some alkanes. The similarity of the structures is clear.

Figure 7.4
The structural formulae of the first six alkanes

```
     H           H   H             H   H   H
     |           |   |             |   |   |
 H - C - H   H - C - C - H     H - C - C - C - H
     |           |   |             |   |   |
     H           H   H             H   H   H
```

methane CH_4 ethane C_2H_6 propane C_3H_8

```
     H   H   H   H             H   H   H   H   H
     |   |   |   |             |   |   |   |   |
 H - C - C - C - C - H     H - C - C - C - C - C - H
     |   |   |   |             |   |   |   |   |
     H   H   H   H             H   H   H   H   H
```

butane C_4H_{10} pentane C_5H_{12}

```
     H   H   H   H   H   H
     |   |   |   |   |   |
 H - C - C - C - C - C - C - H
     |   |   |   |   |   |
     H   H   H   H   H   H
```

hexane C_6H_{14}

The alkanes have no **double bonds**. For this reason they are known as **saturated hydrocarbons**, as no more atoms can bond to the carbons.

The **general equation** for the complete combustion of the alkanes is:

$$C_nH_{2n+2} + \tfrac{1}{2}(3n+1)O_2 \rightarrow nCO_2 + (n+1)H_2O$$

For methane, where n = 1, the formula becomes:

$$CH_4 + 2O_2 \rightarrow CO_2 + 2H_2O$$

And for ethane, where n = 2, the formula becomes:

$$C_2H_6 + 3\tfrac{1}{2}O_2 \rightarrow 2CO_2 + 3H_2O$$

The general equation for the incomplete combustion of alkanes to produce carbon monoxide is:

$$C_nH_{2n+2} + (n+\tfrac{1}{2})O_2 \rightarrow nCO + (n+1)H_2O$$

For methane, where n = 1, the formula becomes:

$$CH_4 + 1\tfrac{1}{2}O_2 \rightarrow CO + 2H_2O$$

And for ethane, where n = 2, the formula becomes:

$$C_2H_6 + 2\tfrac{1}{2}O_2 \rightarrow 2CO + 3H_2O$$

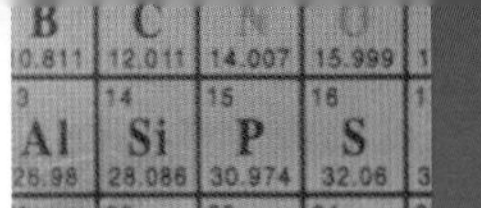

Did you know?

Air is about 20% oxygen so for every millilitre of oxygen required to burn an alkane, 5 millilitres of air are required. For an alkane of formula C_nH_{2n+2}, the amount of oxygen required for the complete combustion of 1 millilitre of the alkane vapour is $\frac{1}{2} \times (3n+1)$. This means that $2\frac{1}{2} \times (3n+1)$ millilitres of air is required.

Alkane	Millilitres of air to burn 1 millilitre of alkane vapour
methane	10
ethane	17.5
propane	25
butane	32.5
pentane	40

To burn 1 millilitre of petrol vapour requires between 60 and 70 millilitres of air. Technical problems make it impossible to run at this ratio. In car engines the mixture has a much higher petrol content than the theoretical value. This means the combustion is incomplete and the products include significant amounts of carbon monoxide. Catalytic converters change most of this back to carbon dioxide before it is released into the atmosphere.

The alkenes

The general formula of the **alkenes** is C_nH_{2n}. Alkanes are flammable and are generally more reactive than alkanes. (See also section 4.4.)

Figure 7.5 *The structural formulae of the first four alkenes*

```
H   H              H   H   H
|   |              |   |   |
C = C – H          C = C – C – H
|                  |       |
H                  H       H

ethene C2H4        propene C3H6

H   H   H   H          H   H   H   H   H
|   |   |   |          |   |   |   |   |
C = C – C – C – H      C = C – C – C – C – H
|       |   |          |       |   |   |
H       H   H          H       H   H   H

butene C4H8            pentene C5H10
```

The similarity of the structures is clear from Figure 7.5.

The alkenes have one double bond. For this reason they are known as **unsaturated hydrocarbons**. They are more reactive than the alkanes because of the double bond.

A double bond is shorter and therefore stronger than a single bond. It may seem strange that it is the double bond which is the part of the molecule that is attacked by reactants. The reason is that the two bonds in a double bond are not identical. The second bond is much weaker than the first. The **bond energy** of a single bond (C–C) is 340 kJ mol^{-1}; for a double bond (C = C) the bond energy is 610 kJ mol^{-1}. Clearly the double bond is stronger than a single bond but the second bond has a bond energy of 610 – 340 = 270 kJ mol^{-1}. The second bond has only about 80% of the strength of the first bond.

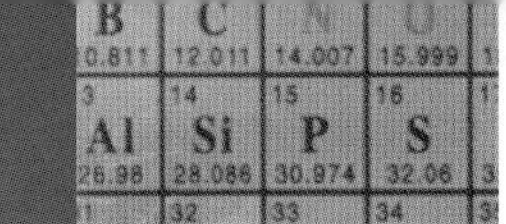

All the reactions of the alkenes (apart from combustion) are **addition reactions** where some substance attacks the double bond. There are three important reactions.

1 The addition of hydrogen to form an alkane

Using the example of ethene:

$$C_2H_4 + H_2 \rightarrow C_2H_6$$

This reaction is used commercially to convert vegetable oils to margarine. Vegetable oils are liquids. They contain unsaturated chains of carbon atoms. Reacting the oil with hydrogen (using a suitable catalyst) will cause some of the double bonds to be **hydrogenated**. This process converts the oil to a solid.

Figure 7.6
Liquid vegetable oil is hydrogenated to produce solid margarine

Did you know?

Unsaturated fats (like vegetable oils) are less harmful to your health than saturated fats (animal fats). Converting unsaturated vegetable oils to saturated fats to make margarine increases the danger to health.

Food labels indicate what percentage of the fat they contain is saturated fat and what percentage is unsaturated fat.

2 The addition of bromine to form a dibromoalkane

Again using the example of ethene:

$C_2H_4 + Br_2(aq) \rightarrow C_2H_4Br_2$ (more usually written as) $CH_2Br.CH_2Br$

This reaction can be used to tell the difference between alkanes and alkenes. An alkane will not react with bromine water (bromine dissolved in water) but an alkene will. If bromine water is shaken with an alkane, the bromine water remains brown but if shaken with an alkene it is decolourised.

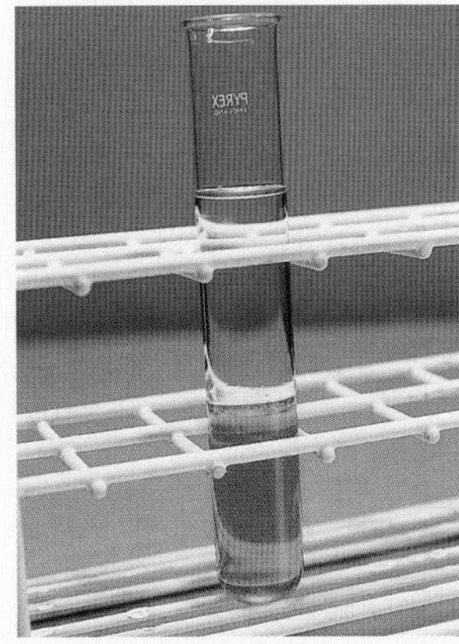

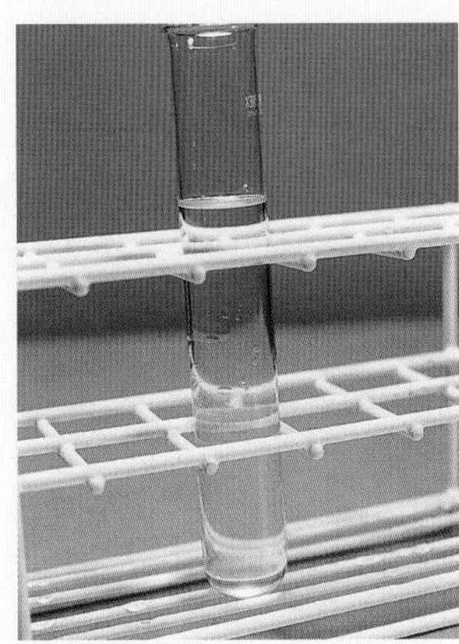

Figure 7.7
When an alkene is shaken with bromine water, it will decolourise it

Figure 7.8 shows how the attacking substance opens up the double bond. You can see from this diagram why dibromoethane is written as $CH_2Br.CH_2Br$ instead of $C_2H_4Br_2$.

Figure 7.8
The addition reactions of alkenes

ethene + hydrogen → ethane

$$H_2C{=}CH_2 + H{-}H \rightarrow H_3C{-}CH_3$$

ethene + bromine → dibromoethane

$$H_2C{=}CH_2 + Br{-}Br \rightarrow CH_2Br{-}CH_2Br$$

Did you know?

The main reason why dibromoethane is written as $CH_2Br.CH_2Br$ instead of $C_2H_4Br_2$ is that there are two dibromoethanes. One with a bromine on each carbon atom; the other with both bromine atoms on the same carbon atom. They are written as $CH_2Br.CH_2Br$ and $CHBr_2.CH_3$ to tell them apart.

They also have different names. $CHBr_2.CH_3$ is called 1,1-dibromoethane (because both bromine atoms are on the same carbon atom) and $CH_2Br.CH_2Br$ is called 1,2-dibromethane (because one bromine atom is on the 1st carbon atom and the other is on the 2nd carbon atom.)

Did you know?

Hydrocarbons containing halogen atoms are used as anaesthetics. One of the earliest anaesthetics was trichloromethane, $CHCl_3$, commonly known as chloroform. More recent anaesthetics include halothane, sometimes called flurothane, $CF_3.CHBrCl$. One of the advantages of these anaesthetics is that they are non-flammable.

3 Addition with itself to form a polymer

This is dealt with later (section 7.8).

The combustion of alkenes is not an addition reaction. Combustion results in the whole molecule being destroyed.

The general equation for the complete combustion of the alkenes is:

$$C_nH_{2n} + 1\tfrac{1}{2}nO_2 \rightarrow nCO_2 + nH_2O$$

And for ethene, where n = 2, the formula becomes:

$$C_2H_4 + 3O_2 \rightarrow 2CO_2 + 2H_2O$$

The general equation for the incomplete combustion of alkenes is:

$$C_nH_{2n} + nO_2 \rightarrow nCO + nH_2O$$

And for ethene, where n = 2, the formula becomes:

$$C_2H_4 + 2O_2 \rightarrow 2CO + 2H_2O$$

Summary

- An **homologous series** is a family of compounds with similar structures and chemical properties.
- Alkanes, alkenes, alcohols and carboxylic acids are all examples of homologous series.
- **Alkanes** have the general formula C_nH_{2n+2}.
- Alkanes are **saturated hydrocarbons** because they only have C–C single bonds.
- Alkanes are flammable but are otherwise unreactive.
- **Alkenes** have the general formula C_nH_{2n}.
- Alkenes are **unsaturated hydrocarbons** because they contain a C = C double bond.
- Alkenes are flammable and are more reactive than alkanes. They will react with hydrogen, bromine and water (steam). In each case the double bond is attacked.
- Alkenes can **polymerise**.

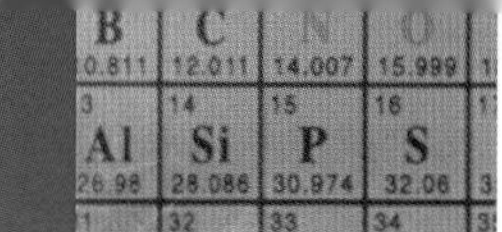

7.4	
Co-ordinated	Modular
10.4	21 (14.7)

Isomerism

The alkanes in Figure 7.4 are all '**straight chain**' alkanes. Not all alkanes are straight chains, some exist as '**branched chains**'. Figure 7.9 shows two alkanes. One is a straight chain alkane (butane), the other is a branched chain alkane. The branched chain alkane has the same formula as butane but a different structure. It also has different physical properties from butane. It is called 2-methylpropane to distinguish it from butane.

Figure 7.9
The isomers of butane

```
    H   H   H   H              H   H   H
    |   |   |   |              |   |   |
H - C - C - C - C - H      H - C - C - C - H
    |   |   |   |              |   |   |
    H   H   H   H              H   |   H
                                   |
                               H - C - H
                                   |
                                   H
```

butane C_4H_{10}

2-methylpropane
(isomer of butane C_4H_{10})

Organic compounds with the same **molecular formula** (and same **relative molecular mass**) are called **isomers.** Butane (C_4H_{10}) and 2-methylpropane (C_4H_{10}) are isomers.

Naming isomers

The names of isomers are worked out using the following rules:

Rule for naming isomer	Rule applied to the isomer of butane
1 a) Find the longest chain of carbon atoms.	3 C atoms
b) What is the name of the alkane with this number of carbon atoms?	propane
2 To which carbon atom is/are the branch(es) bonded? Count from the end to make the number/s as small as possible.	2
3 Name the branch by dropping the '–ane' ending and replacing it by '–yl'.	1 carbon atom (would be methane) so methyl
4 Put the parts of the name together in the sequence Rule 2, Rule 3, Rule 1. Put a hyphen between the number and the name. Join the two parts of the name to make one word.	2-methylpropane

The isomers of pentane

Figure 7.10 show the three isomers of pentane. Check to see how the rules are applied in this example.

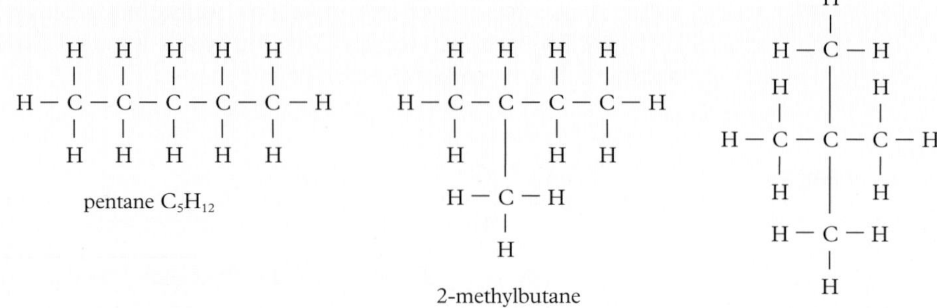

Figure 7.10 *The structural formulae of the isomers of pentane*

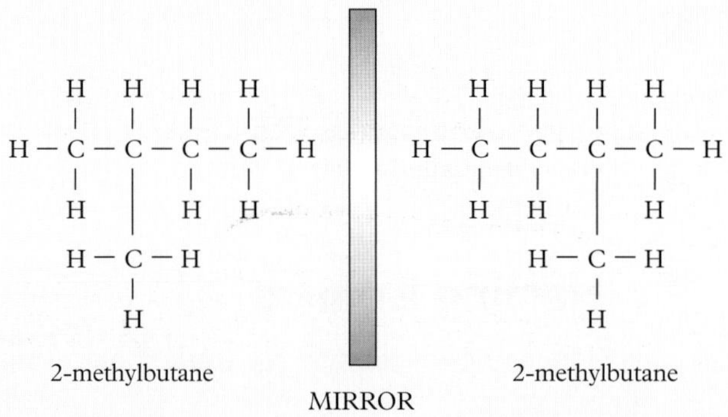

Figure 7.11 *Why there is not another isomer of pentane*

It is easy to think that there might be another isomer of pentane. Figure 7.11 shows why this is not the case. The isomer that looks as if it ought to be 3-methylpropane is just a mirror image of 2-methylpropane. When you are looking for isomers you must take care not to make this mistake. It is easy to make this mistake because structural formulae are two-dimensional and the molecules are three-dimensional. Figure 7.12 is a more accurate picture of the true shape of the three isomers of pentane.

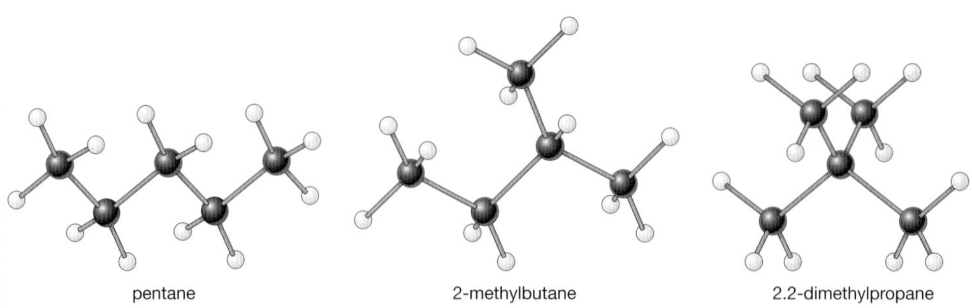

Figure 7.12 *Ball and stick representations of the isomers of pentane*

Did you know?

The number of isomers increases as the number of carbon atoms increases, but there is no mathematical formula to work out the number of isomers of a particular alkane.

There are 5 isomers of C_6H_{14}

18 isomers of C_8H_{18}

75 isomers of $C_{10}H_{22}$

1858 isomers of $C_{14}H_{30}$

60 523 isomers of $C_{18}H_{38}$

Summary

- **Isomers** are substances with the same molecular formula but different chemical structures.
- Isomers have different physical (and often chemical) properties.

7.5 The physical properties of alkanes

Co-ordinated	Modular
10.4	21 (14.7)

The **melting points** and **boiling points** of the alkanes increase as the number of carbon atoms increases.

Figure 7.13
Table of melting points and boiling points for the first 10 alkanes

Name of alkane	Number of carbon atoms	Melting point / °C	Boiling point / °C
methane	1	−182	−164
ethane	2	−183	−89
propane	3	−190	−42
butane	4	−138	−1
pentane	5	−130	36
hexane	6	−95	69
heptane	7	−91	98
octane	8	−57	126
nonane	9	−51	151
decane	10	−30	174

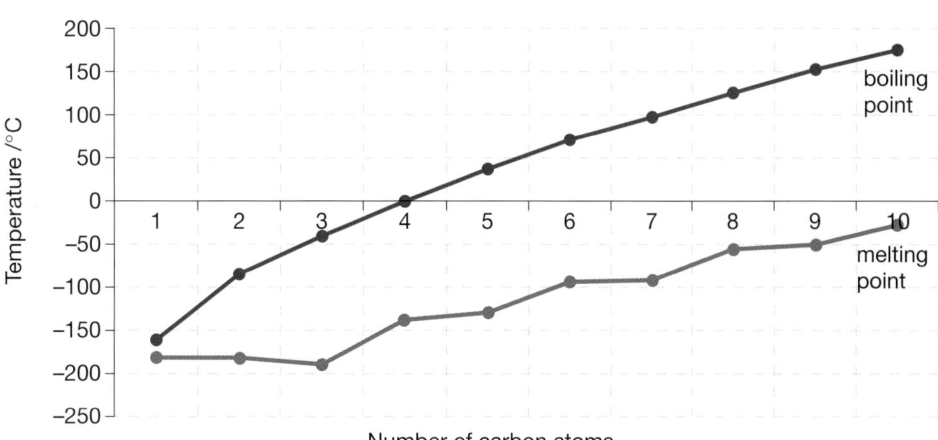

Figure 7.14
Graph of melting points and boiling points for the first 10 alkanes

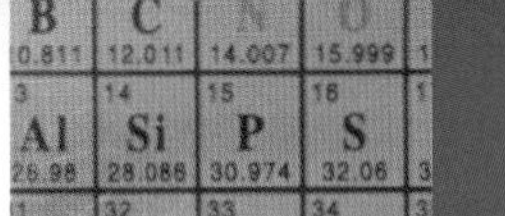

The melting points have an irregular trend but the boiling points show a very clear pattern of behaviour.

Alkane molecules have a slight attraction for each other. Weak forces hold the molecules together. The longer the molecule, the more contact there is between molecules and the stronger the forces holding the molecules together (see Figure 7.15). This means that more energy is needed to overcome these forces and change the alkane into a gas. So the longer the molecule, the higher the boiling point.

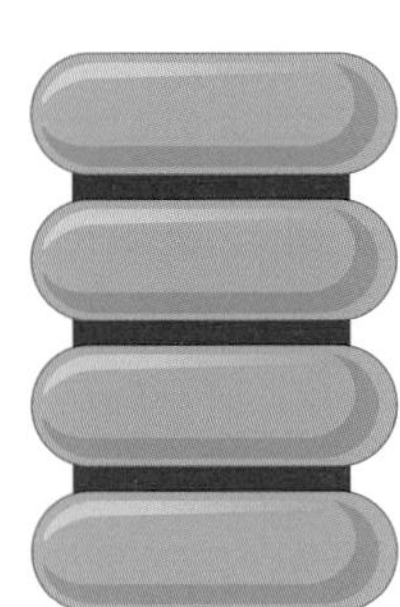

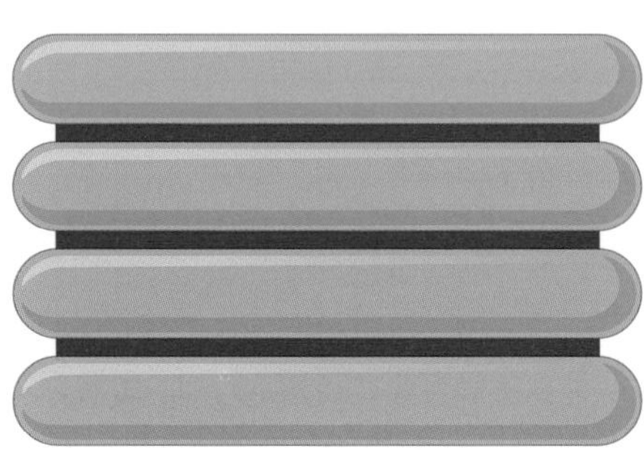

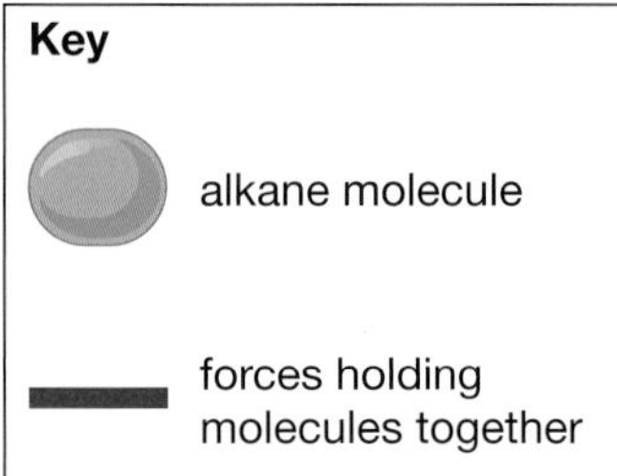

Figure 7.15
How the length of alkane molecules affects the forces holding them together

Physical properties of isomers

Although the chemical properties of the isomers of alkanes are quite similar, their physical properties are not. Figure 7.16 shows the boiling points of the isomers of butane and pentane.

Figure 7.16
The boiling points of the isomers of butane and pentane

Alkane	Boiling point / °C
butane	−1
2-methylpropane	−12
pentane	36
2-methylbutane	28
2,2-dimethylpropane	10

Branched chain alkanes have lower boiling points than their straight chain isomer. The more branched a chain is, the lower its boiling point. Figure 7.12 shows how the structures of the isomers of pentane differ. The molecules of branched chain alkanes cannot get as close together as those of straight chain alkanes. This means the forces holding the molecules together are weaker; less energy is needed to separate the molecules so the boiling point of the compound will be lower.

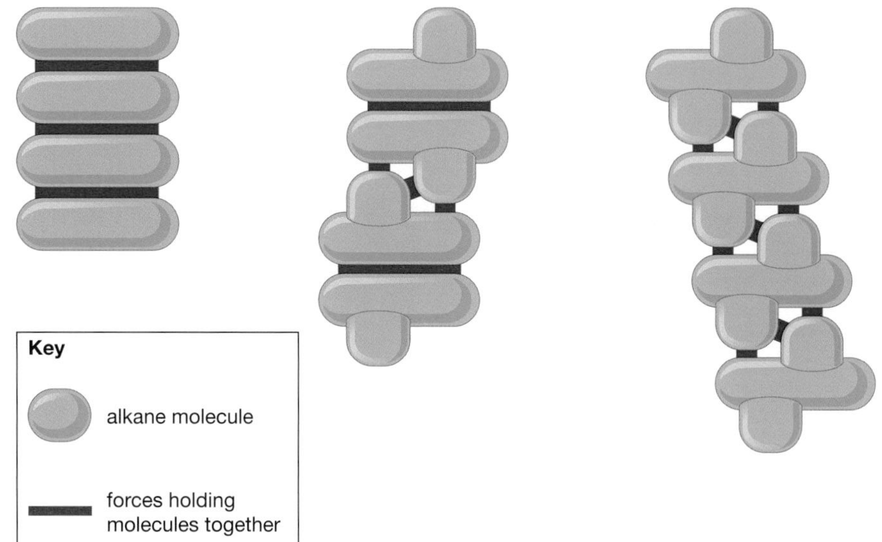

Figure 7.17
How branching in alkane molecules affects the forces holding them together

Topic questions

1 a) What is the general formula of an alkene?
   b) What is the chemical equation for the complete combustion of ethene?

2 a) Write a word equation for:
      i) the complete combustion of methane.
      ii) the incomplete combustion of propene to produce carbon monoxide.
   b) Write a balanced chemical equation for each of these word equations.

3 Explain, using bond energies, why alkenes are more reactive than alkanes.

4 Give **three** examples of addition reactions of alkenes.

5 Name the five isomers of hexane and draw the structural formulae.

6 Explain why the boiling point of pentane is:
   a) higher than that of butane
   b) higher than that of 2-methylbutane.

Summary

- The melting points of alkanes generally increase as the number of carbon atoms increases.
- The boiling points of alkanes increase in a regular way as the number of carbon atoms increases.

7.6 Alcohols

Co-ordinated	Modular
10.4	21 (14.8)

Ethanol

Ethanol is the **alcohol** that is present in alcoholic drinks. It is made by the **fermentation** of sugars in the presence of yeast.

$$C_6H_{12}O_6 \rightarrow 2C_2H_5OH + 2CO_2$$

glucose → ethanol + carbon dioxide

Yeast produces an **enzyme** (zymase) which is a **biological catalyst** for the reaction. In the fermentation process, the raw materials are mixed with water and yeast and left at a temperature slightly higher than room temperature. The reaction is **anaerobic** so air is excluded.

Figure 7.18 shows a 'home brewing' apparatus. The airlock in the bung prevents air getting into the bottle. The ethanol produced poisons the yeast and eventually the reaction stops. Once the reaction has stopped, the ethanol can be removed from the mixture by fractional distillation.

Figure 7.18 *Home brewing apparatus*

Did you know?

Spirits like whisky and brandy are made by the fractional distillation of wines and ales. The first substance distilled off is methanol (boiling point 64°C). This is a toxic alcohol so must be removed from the mixture. (It is sold for industrial use.) The second substance is ethanol (boiling point 78°C). This is kept. Finally the water is distilled off and discarded. The flavours and colours of the original material remain and these are blended with the ethanol to produce spirits.

Ethanol is used as a solvent and as a fuel. In some third world countries where the cost of oil is high, ethanol, made by fermenting crops like sugar cane, is added to petrol to produce a fuel called gasohol. Ethanol is a good fuel because it burns readily with a blue, non-smoky flame.

In industry, ethanol is also made from ethene.

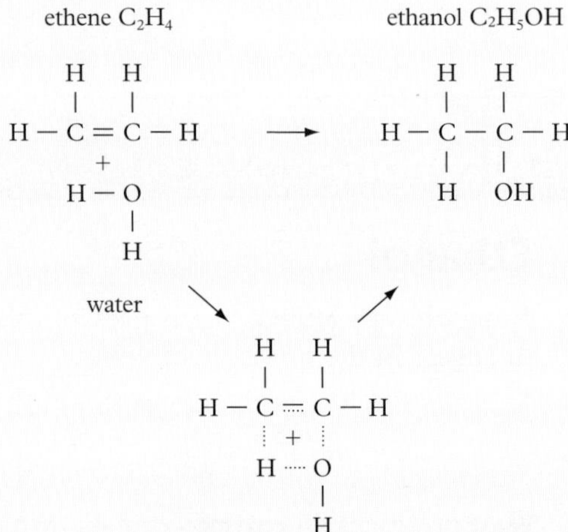

The equation shows how water (as steam) can be added to the double bond. The process requires temperatures of about 300°C and pressures of about 65 atmospheres (6500 kPa). Phosphoric acid is used as a catalyst.

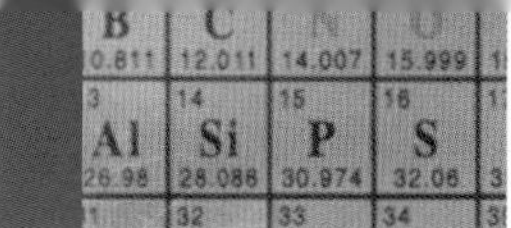

Advantages and disadvantages of the two methods of making alcohol

When there are two methods of producing a chemical, there is competition between these methods. Usually the cheaper method is favoured. This is why the industrial manufacture of chemicals always uses the cheapest possible raw material. If two methods are available and both are used it is because there are advantages and disadvantages in each method so neither is the 'winner'.

The table below analyses the advantages and disadvantages of each method of making ethanol.

Figure 7.19 *Comparing the advantages and disadvantages of the two methods of preparing ethanol*

Factor considered	Method	
	Fermentation	Ethene + water
Rate of production (rate of reaction)	Comparatively slow. Takes some weeks for the mixture to ferment fully.	Relatively fast
Quality of the product	After fractional distillation is about 96% pure ethanol. (The impurity is water.)	Not very pure. The product is not suitable for consumption or medical purposes.
Manufacturing process	Produced in batches. Batch processes are not very efficient and industry tries to avoid them.	Continuous production. Reactants enter one end of the plant and products emerge from the other continuously.
Use of resources	Uses sustainable resources. Yeast is a living organism and reproduces. Plant material used can be grown.	Uses finite resources. Ethene is obtained from oil.

The cells shaded grey are for the process which has the advantage.

No reference is made in the table to the energy costs. The fermentation process uses very little energy until the fractional distillation stage. The other process uses energy to produce steam to heat the reaction vessel to a high temperature and to pressurise the reactants.

The homologous series of the alcohols

Alcohols form an homologous series. The general formula is $C_nH_{2n+1}OH$. Notice that the H symbols are not all together. This is to show that ethanol has an –OH group attached to it. The –OH group is an example of a **functional group**. Substances with the –OH functional group have similar chemical reactions because of that group.

Alcohols are named systematically. The alkane with one carbon atom is called methane and the alcohol with one carbon atom is called methanol.

```
     H              H   H             H   H   H
     |              |   |             |   |   |
 H — C — OH     H — C — C — OH    H — C — C — C — OH
     |              |   |             |   |   |
     H              H   H             H   H   H
```

methanol CH_3OH ethanol C_2H_5OH propanol C_3H_7OH

Figure 7.20 *Some members of the homologous series of alcohols*

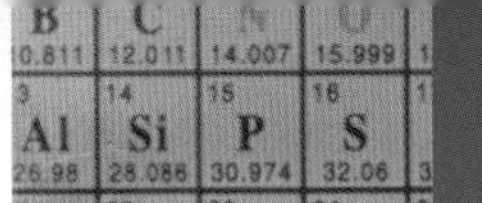

Did you know?

Isomerism also occurs in alcohols.

```
      H   H   H              H   H   H
      |   |   |              |   |   |
  H — C — C — C — H      H — C — C — C — H
      |   |   |              |   |   |
      H   H   OH             H   OH  H
```

In each case the chains are straight chains but the functional –OH group is situated on a different carbon atom. The chemical and physical properties of these two alcohols will be slightly different.

The naming system is similar to that used in naming branched chain alkanes. They are called propan-1-ol and propan-2-ol, respectively.

The **steroid**, cholesterol, contains the –OH functional group. Cholesterol is an essential steroid for humans but too much of it can cause heart disease.

The chemical reactions of alcohols

All alcohols are flammable. The products of complete combustion are carbon dioxide and water (steam). For ethanol the reaction is:

$$C_2H_5OH + 3O_2 \rightarrow 3H_2O + 2CO_2$$

Alcohols also react with sodium to form hydrogen. This reaction occurs because of the presence of the –OH functional group. The reaction shown is for ethanol but it also works equally well with other alcohols.

$$2Na + 2C_2H_5OH \rightarrow 2C_2H_5ONa + H_2$$

The product C_2H_5ONa is called sodium ethoxide. It is an ionic compound. This is the first **ionic** compound mentioned in this chapter. This is because the bonds in organic compounds are almost entirely **covalent**.

Summary

- **Alcohols** contain the –OH **functional group**.
- **Ethanol** can be made by the **fermentation** of sugar by the enzymes in **yeast** or by the addition of water (steam) across the double bond of ethene. There are advantages and disadvantages with either method of preparation.
- Ethanol reacts with sodium to form sodium ethoxide. Other alcohols react with sodium in a similar way.

Topic questions

1 Which of the following is the formula of ethanol?
   a) CH_3OH
   b) C_2H_4
   c) C_2H_5OH
   d) C_2H_6

2 In some parts of the world alcohol is added to petrol to make a fuel for vehicles. Which method of alcohol manufacture is used? Explain why this method is used.

3 Write a balanced chemical equation for the reaction of sodium with methanol. Name the products of the reaction.

7.7	
Co-ordinated	Modular
10.4	21 (14.9)

Carboxylic acids

Carboxylic acids also form an homologous series. Figure 7.21 shows some of the members of that series.

methanoic acid	ethanoic acid	propanoic acid
H.COOH	CH_3.COOH	C_2H_5.COOH

Figure 7.21 *The first three members of the homologous series of carboxylic acids*

In carboxylic acids the functional group is –COOH.

The reactions of carboxylic acids

All carboxylic acids are weak acids (see Chapter 9). This is because the –O–H bond in the –COOH functional group is slightly ionised. They have typical acidic behaviours. They are neutralised by alkalis and they react with carbonates and hydrogencarbonates to produce carbon dioxide. In each of these reactions a carboxylic acid salt is produced.

For example:

$$H.COOH + NaOH \rightarrow H_2O + H.COONa$$

methanoic acid + sodium hydroxide → water + sodium methanoate

$$2CH_3.COOH + Na_2CO_3 \rightarrow CO_2 + H_2O + 2CH_3.COONa$$

ethanoic acid + sodium carbonate → carbon dioxide + water + sodium ethanoate

$$C_2H_5.COOH + NaHCO_3 \rightarrow CO_2 + H_2O + C_2H_5.COONa$$

propanoic acid + sodium hydrogencarbonate → carbon dioxide + water + sodium propanoate

The salts produced are ionic, so C_2H_5.COONa could be written $C_2H_5.COO^- Na^+$.

Notice that the formulae for salts of carboxylic acids are written 'backwards'. The **anion** (negative ion) is written first followed by the **cation** (positive ion). This is the conventional way of writing the formulae of organic salts. But there is nothing wrong in writing them with the positive ion first (e.g. $Na^+C_2H_5.COO^-$).

In carboxylic acids it is only the hydrogen in the –COOH functional group that is replaced. This is because the OH bond in the –COOH functional group is partly ionic but the –C–H bonds are all covalent.

Carboxylic acids will react with alcohols. The products are called **esters**. The reaction is reversible and very slow. Concentrated sulphuric acid is used as a catalyst for this reaction.

$$CH_3.COOH + C_2H_5OH \rightleftharpoons CH_3.COO.C_2H_5 + H_2O$$

ethanoic acid + ethanol ⇌ ethyl ethanoate + water

Although this reaction seems similar to the reaction between a carboxylic acid and an alkali, it is wrong to think of the reaction like this. Alcohols are not at all like alkalis and esters do not behave like salts. All salts have ions; all esters are covalently bonded. Esters have characteristic odours and flavours and are used as fragrances and food flavourings.

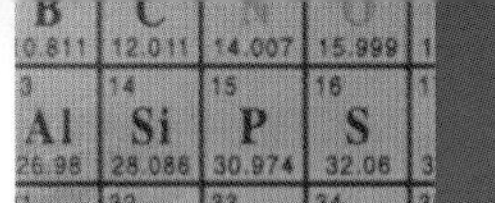

Figure 7.22
All these things contain common carboxylic acids

Did you know?

The scent of flowers and the flavour of fruits are due to mixtures of esters. Ethyl ethanoate has the fragrance of apples, ethyl butanoate smells of pineapples and octyl ethanoate smells of oranges.

Uses of carboxylic acids

Ethanoic acid (probably the first acid ever used by man) is obtained by the oxidation of ethanol. It is the acid present in vinegar and in 'sour' alcoholic drinks. Ethanoic acid is also used in the manufacture of the fibre, acetate rayon.

Other common carboxylic acids include:

- citric acid, present in citrus fruits like oranges, lemons, limes and grapefruit and used as an additive in many soft drinks
- ethanoylsalicylic acid (better known as 'aspirin') used in the relief of pain and to 'thin' the blood of people at risk of heart attacks
- ascorbic acid (vitamin C) present in fresh fruit and vegetables.

Did you know?

Methanoic acid is found in the 'sting' of stinging nettles and the bite of ants. The old name for methanoic acid was formic acid from the Latin word for ant.

Topic questions

1 Which functional group is present in all carboxylic acids?

2 Give **three** examples of reactions where carboxylic acids behave as typical acids. Write a balanced chemical equation to illustrate each reaction.

3 a) Write a chemical equation for the reaction of propanoic acid with methanol. Name the products of the reaction.
  b) The organic compound produced in a) is an isomer of ethyl ethanoate.
    i) What is meant by the word 'isomer'?
    ii) Use diagrams of the structures to explain why the organic compound produced in a) and ethyl ethanoate are different compounds.

Summary

- **Carboxylic acids** contain the –COOH functional group.
- Carboxylic acids are weak acids. They have typical acid behaviour – they can be neutralised by alkalis and react with carbonates and hydrogencarbonates to produce carbon dioxide.
- Carboxylic acids react with alcohols to form **esters**. This is a **reversible reaction**.

7.8	
Co-ordinated	Modular
10.4	21 (14.6/14.9)

Polymers

The structure of polymers

Polymers are the substances generally called **plastics**. They are made of long molecules in the form of chains. The atoms in these molecules are held together with strong covalent bonds. The molecules are tangled together and joined to each other by **cross-linking bonds**.

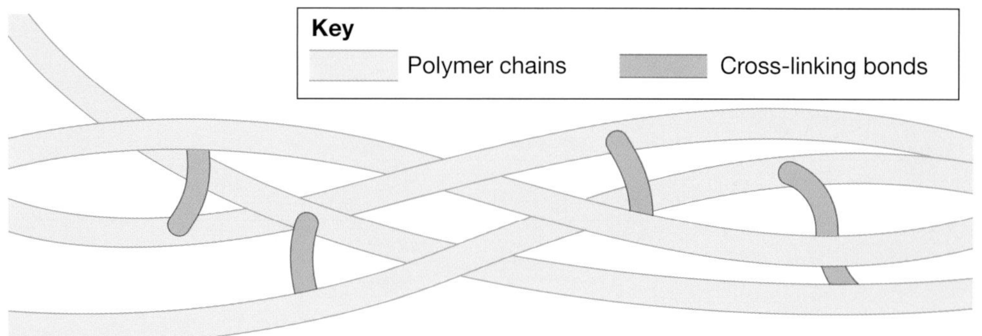

Figure 7.23
The arrangement of molecules in a polymer

Formation of polymers

Most polymers are made from compounds containing the – C = C – bond These compounds can link together because of the double bond. The process of linking these bonds together is called **addition polymerisation**. The individual molecules that join together are called **monomers**. So molecules of the monomer ethene will link together to form the polymer poly(ethene) (often called polythene) and the monomer propene will form the polymer poly(propene) (often called polypropylene). The monomer chloroethene ($CH_2{=}CHCl$) polymerises to poly(chloroethene) usually called polyvinylchloride or PVC.

The equation for the polymerisation of ethene is:

$$n\mathrm{CH_2{=}CH_2} \rightarrow [\mathrm{-CH_2-CH_2-}]_n$$

Figure 1.24 shows the way ethene polymerises.

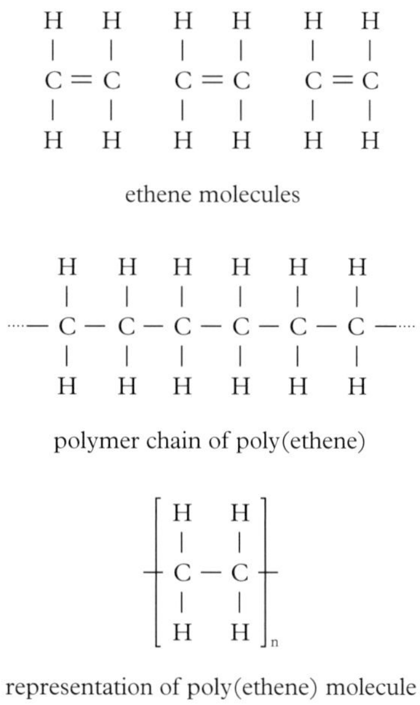

$$\begin{array}{ccc} \mathrm{H}\quad\mathrm{H} & \mathrm{H}\quad\mathrm{H} & \mathrm{H}\quad\mathrm{H} \\ |\quad\;\;| & |\quad\;\;| & |\quad\;\;| \\ \mathrm{C}=\mathrm{C} & \mathrm{C}=\mathrm{C} & \mathrm{C}=\mathrm{C} \\ |\quad\;\;| & |\quad\;\;| & |\quad\;\;| \\ \mathrm{H}\quad\mathrm{H} & \mathrm{H}\quad\mathrm{H} & \mathrm{H}\quad\mathrm{H} \end{array}$$

ethene molecules

$$\begin{array}{c} \mathrm{H}\quad\mathrm{H}\quad\mathrm{H}\quad\mathrm{H}\quad\mathrm{H}\quad\mathrm{H} \\ |\quad\;\;|\quad\;\;|\quad\;\;|\quad\;\;|\quad\;\;| \\ \cdots-\mathrm{C}-\mathrm{C}-\mathrm{C}-\mathrm{C}-\mathrm{C}-\mathrm{C}-\cdots \\ |\quad\;\;|\quad\;\;|\quad\;\;|\quad\;\;|\quad\;\;| \\ \mathrm{H}\quad\mathrm{H}\quad\mathrm{H}\quad\mathrm{H}\quad\mathrm{H}\quad\mathrm{H} \end{array}$$

polymer chain of poly(ethene)

$$\left[\begin{array}{c} \mathrm{H}\quad\mathrm{H} \\ |\quad\;\;| \\ -\mathrm{C}-\mathrm{C}- \\ |\quad\;\;| \\ \mathrm{H}\quad\mathrm{H} \end{array}\right]_n$$

representation of poly(ethene) molecule

Figure 7.24
Polymerisation of ethene to form poly(ethene)

Did you know?

The polymerisation process is quite complex. Poly(ethene) exists in two forms, low density poly(ethene) and high density poly(ethene). The low density form requires a pressure of about 2000 atmospheres and temperature of about 250°C. The high density form is manufactured at atmospheric pressure and a temperature of about 100°C but requires a complex combination of catalysts and is more expensive to produce.

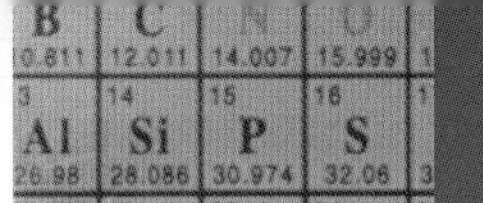

Figure 7.25 shows the polymerisation of chloroethene to form PVC.

Figure 7.25
How monomers of chloroethene link to form PVC

```
H   Cl    H   Cl    H   Cl
|   |     |   |     |   |
C = C     C = C     C = C
|   |     |   |     |   |
H   H     H   H     H   H
```

chloroethene molecules

```
       H   Cl  H   Cl  H   Cl
       |   |   |   |   |   |
.....— C — C — C — C — C — C —.....
       |   |   |   |   |   |
       H   H   H   H   H   H
```

polymer chain of poly(chloroethene)

```
  [ H   Cl ]
  [ |   |  ]
 —[ C — C  ]—
  [ |   |  ]
  [ H   H  ]n
```

representation of poly(chloroethene) molecule

Thermosoftening and thermosetting plastics

Poly(ethene), poly(propene) and PVC are **thermosoftening plastics**. Thermosoftening plastics have weak cross-linking bonds. These bonds are easily broken by heat and the polymer can be reshaped. When it cools, new cross-link bonds form. This means that thermosoftening plastics can easily be moulded into shape.

Thermosetting plastics are polymers in which the cross-linking bonds are formed when the material is heated. These bonds are very much stronger than those in thermosoftening plastics. Heating cannot reshape thermosetting plastics. The plastic Melamine (used in the manufacture of furniture) is an example of a thermosetting polymer. Many glues are thermosetting polymers.

Topic questions

1 The structural formula of propene is:

```
H   H
|   |
C = C     H
|    \   /
H      C
      / \
     H   H
```

Draw the structural formula of poly(propene).

2 Electric switches and sockets are made from plastic.
   a) Why is this safer than making them from metal?
   b) Is the plastic used thermosoftening or thermosetting? Explain why that type of plastic is used.

3 What is the name of the monomer used to make poly(styrene)?

Summary

- **Polymers** are formed when **monomer** alkenes bond together.
- Polymers can be **thermosoftening** or **thermosetting**.

Examination questions

1 Cars in Brazil use ethanol as a fuel instead of petrol (octane). The ethanol is produced by the fermentation of sugar solution from sugar cane.
   a) What must be added to sugar solution to make it ferment? *(1 mark)*
   b) Which is the most suitable temperature for a fermentation.

      0°C   10°C   30°C   70°C   100°C *(1 mark)*
   c) i) What compounds are formed by the complete combustion of ethanol? *(2 marks)*
      ii) Why are these compounds **not** harmful to the environment? *(1 mark)*
   d) Suggest why pollution from cars is less when using ethanol instead of petrol. *(1 mark)*
   e) Give **one** reason why ethanol is **not** used as a fuel for cars in Britain. *(1 mark)*
   f) Some information about octane and ethanol is shown.

Property	Octane	Ethanol
Melting point in °C	−57	−113
Boiling point in °C	125	78.5
Density in g/cm^3	0.70	0.79
Heat produced in kJ/mol	5512	1367

Explain a similarity between octane and ethanol that allows ethanol to be used as a fuel in cars. *(2 marks)*

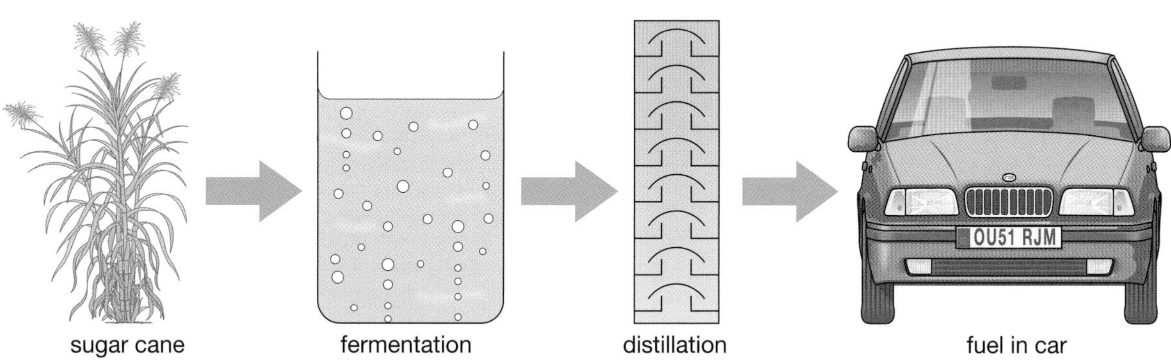

2 a) Use words from the list to copy and complete the passage about organic compounds.

carbon **carbon dioxide** **electricity**
energy **fuels** **neutral** **non-renewable**
renewable **water** **wood**

Some organic compounds are used as ____________ because they release energy when they are burned.
Some of these substances come from fossils. Once used up they cannot be replaced. This means they are ____________.
All organic compounds contain the element ____________. Many also contain the element hydrogen.
When organic compounds containing hydrogen are burned in plentiful supply of air, the two substances formed are ____________ and water. *(4 marks)*

b) Why is it dangerous to burn organic compounds in a limited supply of air? *(2 marks)*

3 Petrol is a fuel used for cars. It is a mixture of hydrocarbons.

a) Name the **two** products formed when a hydrocarbon is burned completely in air. *(2 marks)*

b) i) Name the poisonous gas that is formed when a hydrocarbon is burned in a limited supply of air. *(1 mark)*

ii) Explain why this gas is poisonous. *(2 marks)*

c) Petrol is a fossil fuel and so its supply is limited. Alternative fuels will be needed as it runs out. The table shows data from 1998 for petrol and some alternative fuels.

Fuel	Cost of 100 g (pence)	Energy per 100 g (kJ)	Energy per penny (kJ)
petrol	6.8	4800	706
diesel oil	6.4	4700	734
ethanol	8.5	2900	341
hydrogen	20.0	14300	715
vegetable oil	9.0	3800	422

i) Use the data in the table to explain why diesel oil seems to be a good alternative to petrol. *(1 mark)*

ii) From your knowledge of fuels, give **one** disadvantage of using diesel oil as a replacement fuel for petrol. *(1 mark)*

iii) From the table, hydrogen seems to be a good alternative to petrol. Suggest **one** advantage and **two** disadvantages of using hydrogen as a fuel for cars. *(3 marks)*

4 The table shows some information about alkanes.

Name	Formula	Relative formula mass	Boiling point in °C
methane	CH_4	16	−160
ethane	C_2H_6	30	−90
propane		44	−40
butane	C_4H_{10}	58	
pentane	C_5H_{12}	72	36
hexane	C_6H_{14}	86	68

a) Give the formula of propane. *(1 mark)*

b) i) What happens to the boiling points of the alkanes as the relative formula mass increases? *(1 mark)*

ii) Draw a graph on a grid, with Relative formula mass on horizontal axis and Boiling point (°C) on vertical axis. Plot the points and draw a best fit line. *(3 marks)*

iii) What is the boiling point of butane? *(1 mark)*

iv) Show clearly on the graph how you found the boiling point of butane. *(1 mark)*

c) Which of the following is **not** an alkane.

C_7H_{16} C_9H_{18} $C_{11}H_{24}$ $C_{24}H_{50}$ *(1 mark)*

5 This question is about hydrocarbons.
The table below gives some information about the first ten members of an homologous series. It includes their melting points and boiling points. It was taken from a German Chemistry Textbook.

a) i) Name the homologous series. *(1 mark)*

ii) What is meant by an *homologous series*? *(2 marks)*

iii) Use the information in the table to predict the boiling point of C_9H_{20}. *(1 mark)*

iv) What is the formula of the twelfth member of this series? *(1 mark)*

b) There are **three** hydrocarbons which have the molecular formula C_5H_{12}. The structural formula of one of these is shown opposite (A).

```
    H   H   H   H   H
    |   |   |   |   |
H — C — C — C — C — C — H
    |   |   |   |   |
    H   H   H   H   H
```

(A)

i) Draw the structural formula of the other two compounds (B and C). *(2 marks)*

ii) What name is given to compounds which have the same molecular formula but different structures? *(1 mark)*

iii) Which of the compounds, A, B or C, has the highest boiling point?
Give reasons for your answer. *(3 marks)*

Name	Summenformel	Strukturformelin (Kurzform)	Schmelztemperatur (°C)	Siedetemperatur (°C)
Methan	CH_4	CH_4	−182	−162
Äthan	C_2H_6	CH_3–CH_3	−183	−89
Propan	C_3H_8	CH_3–CH_2–CH_3	−188	−42
Butan	C_4H_{10}	CH_3–$(CH_2)_2$–CH_3	−138	0
Pentan	C_5H_{12}	CH_3–$(CH_2)_3$–CH_3	−130	+36
Hexan	C_6H_{14}	CH_3–$(CH_2)_4$–CH_3	−95	+69
Heptan	C_7H_{16}	CH_3–$(CH_2)_5$–CH_3	−90	+98
Octan	C_8H_{18}	CH_3–$(CH_2)_6$–CH_3	−57	+126
Nonan	C_9H_{20}	CH_3–$(CH_2)_7$–CH_3	−54	
Decan	$C_{10}H_{22}$	CH_3–$(CH_2)_8$–CH_3	−30	+174

Chapter 8

Industrial processes

Key terms alloy • anode (positive electrode) • anodising • blast furnace • cast iron • catalyst • cathode • contact process • corrosion resistant • dehydrating agent • density • electrolysis • electrolyte • electroplating • exothermic • mild steel • neutralising • non-metal • oxidise • reactivity series • redox reaction • reversible reaction • rutile • slag • smelting • steel • transition metal • water of crystallisation • wrought iron

8.1	
Co-ordinated	Modular
10.7	22 (15.1)

Sulphuric acid

Manufacture

Sulphuric acid is made from sulphur, air and water. These substances are relatively inexpensive so the process of manufacturing sulphuric acid is cost effective.

Did you know?

In America, sulphur deposits are found about 200 metres below the surface. Conventional mining would be difficult because the deposits are beneath quicksand. The sulphur is obtained by the Frasch process.

In this process, pressurised water at 155°C is pumped down a 15 cm diameter tube into the ground. The hot water melts the sulphur in the rock and forces it back up the pipe. Compressed air also helps force the sulphur to the surface. The hot water goes down the outside tube ensuring that the inside of the pipe keeps hot so the rising sulphur does not solidify in the pipe. Sulphur produced by this method is 99.5% pure.

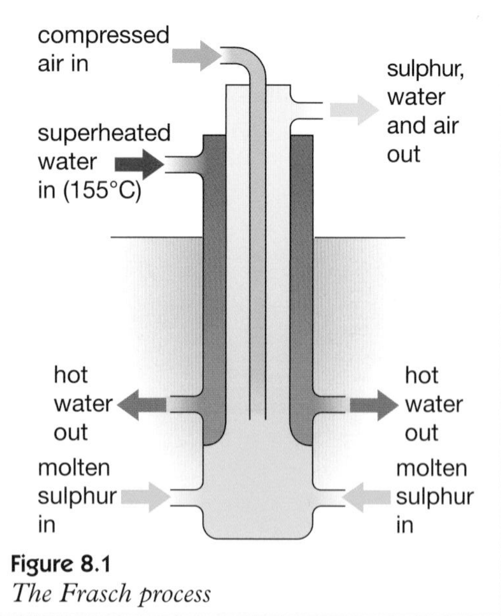

Figure 8.1
The Frasch process

Did you know?

Cost is an important consideration in industrial processes. The capital outlay to build industrial chemical plants is huge. If the running costs are high, the product is comparatively expensive. This means that the invention of a cheaper method will make the original process too expensive and the large investment will be wasted. Processes that use cheap ingredients are unlikely to be undercut. This principle is used in the manufacture of ammonia (from air and water), nitric acid (from ammonia, air and water), sodium hydroxide, chlorine and hydrochloric acid (from sodium chloride and water) and sulphuric acid (from sulphur, air and water).

Usually industrial chemical plants do not produce much pollution. This is partly for economic reasons. Wasting materials costs money so as much as possible is recycled.

Most of the world's sulphuric acid is made by the **contact process**. The process is shown in Figure 8.2. In this process molten sulphur is burned in excess air. Sulphur dioxide (SO_2) is produced in the following reaction.

$$S + O_2 \rightarrow SO_2$$

The sulphur dioxide and any remaining air pass into a converter where more air is added and the sulphur dioxide is oxidised to sulphur trioxide (SO_3). This is a **reversible reaction**. The reaction requires a **catalyst** and vanadium oxide (V_2O_5) is used. A pressure of up to about 2 atmospheres is used.

$$2SO_2 + O_2 \rightleftharpoons 2SO_3$$

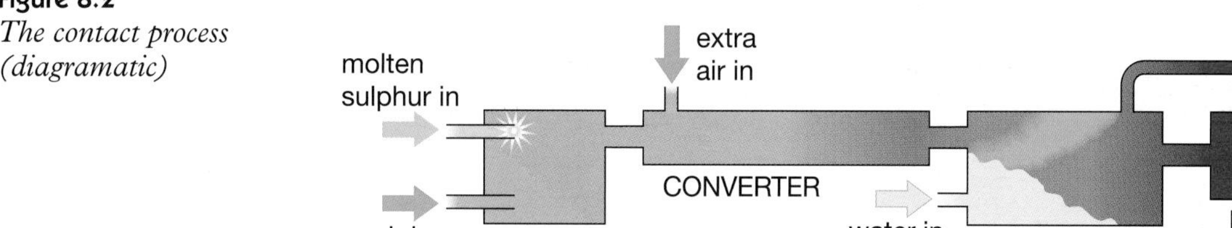

Figure 8.2
The contact process (diagramatic)

The process is strongly **exothermic**. This means that the maximum yield of sulphur trioxide should occur at low temperatures. Unfortunately, even with a catalyst, the reaction is too slow and temperatures of about 450°C are required. From the equation, three 'volumes' of reactants (two volumes of SO_2 and one volume of O_2) produce two volumes of product (SO_3), so increasing the pressure should produce more product. In fact the reaction goes almost to completion even at atmospheric pressure so high pressure (which means extra costs) is not used. The addition of extra air increases the amount of oxygen present. Excess oxygen increases the amount of product.

Figure 8.3
The contact process takes place in plants like this one in Billingham in Cleveland, UK

Theoretically the last stage of the manufacture is to react the sulphur trioxide with water.

$$SO_3 + H_2O \rightarrow H_2SO_4$$

However it is not possible to get sulphur trioxide to react directly with water. This is because the reaction is also very exothermic and produces a fine mist of concentrated acid, which is difficult to control. Instead the sulphur trioxide is dissolved in concentrated sulphuric acid to produce fuming sulphuric acid – sometimes called oleum. The fuming sulphuric acid is carefully diluted with water to produce concentrated sulphuric acid. The acid produced is about 98% pure; the remaining 2% is water.

This is another very exothermic process.

Uses of sulphuric acid

Sulphuric acid is the acid used in car batteries. It is also used in the manufacture of fertilisers and detergents (see Figure 8.4)

Figure 8.4
All these products were made using sulphuric acid as a raw material

Sulphuric acid as a dehydrating agent

A **dehydrating agent** is a substance that will remove water. Concentrated sulphuric acid is a powerful dehydrating agent. It is so powerful that it will even remove the elements of water from a compound.

If concentrated sulphuric acid is added to sugar, the sugar is dehydrated and carbon is produced. This reaction is shown in Figure 8.5.

Figure 8.5
The reaction between concentrated sulphuric acid and sugar

$$C_{12}H_{22}O_{11} \xrightarrow{\text{conc. } H_2SO_4} 12C + 11H_2O$$

Concentrated sulphuric acid will also dehydrate copper(II) sulphate crystals by removing the **water of crystallisation**. When this happens the blue crystals turn white (see Figure 2.6).

$$CuSO_4.5H_2O \xrightarrow{\text{conc. } H_2SO_4} CuSO_4 + 5H_2O$$

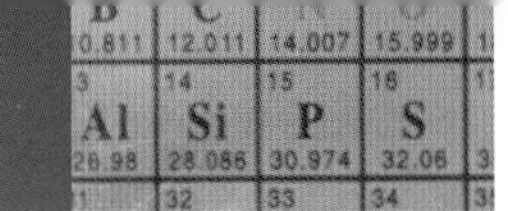

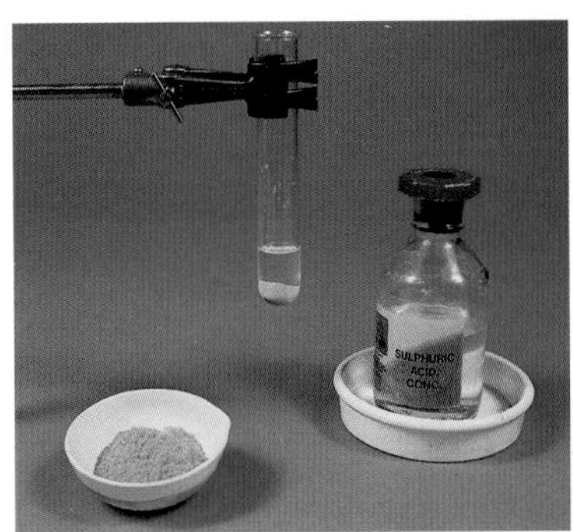

Figure 8.6
The reaction between concentrated sulphuric acid and copper(II) sulphate

Topic questions

1 Name the raw materials used in the manufacture of sulphuric acid.

2 What is the name of the process used to manufacture sulphuric acid?

3 The manufacture of sulphuric acid uses a temperature of 450°C. What would be the advantages and disadvantages of using:
a) 250°C   b) 650°C?

4 Give **three** uses for sulphuric acid.

5 Give an example of a reaction where concentrated sulphuric acid acts as a dehydrating agent.

Summary

- Sulphuric acid is made from the readily available, inexpensive materials sulphur, air and water.
- Sulphuric acid is made by the **contact process**.
- Sulphuric acid is used in the manufacture of car batteries, fertilisers and detergents.
- Sulphuric acid is a **dehydrating agent**. It will dehydrate carbohydrates like sugar to produce carbon. It will also dehydrate blue copper(II) sulphate crystals producing white anhydrous copper(II) sulphate.

8.2	
Co-ordinated	Modular
10.7	22 (15.2)

Aluminium

Aluminium is above iron in the reactivity series. This means that aluminium should corrode quite quickly. In fact aluminium is resistant to corrosion. When first exposed to air, the surface of the metal rapidly **oxidises** but the layer of aluminium oxide formed seals the surface and prevents any further attack. If the oxide layer is removed, the aluminium will react quite rapidly (see Figure 8.7).

Figure 8.7
When the oxide coating of aluminium has been removed, aluminium reacts rapidly with

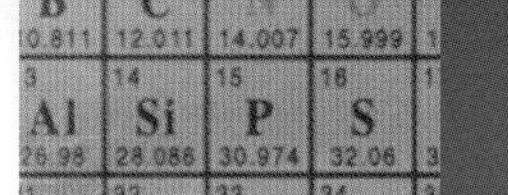

Figure 8.8
Aluminium is anodised to thicken its protective oxide coating

For some applications a thicker protective layer is needed. The process of **anodising** (Figure 8.8) can produce this thicker layer. In the anodising process the existing aluminium oxide layer is removed with sodium hydroxide solution. The aluminium object is then placed in dilute sulphuric acid and made into the **anode (positive electrode)**. Oxygen forms on the surface of the aluminium and a much thicker oxide layer is produced.

Did you know?

Aluminium oxide dissolves in sodium hydroxide because it behaves like an acidic oxide! This is very unusual as acidic oxides are usually non-metal oxides.

$$Al_2O_3 + 2NaOH \rightarrow 2NaAlO_2 + H_2O$$

The substance $NaAlO_2$ is called sodium aluminate.

Aluminium oxide will also behave like a basic oxide. It can be made to dissolve in some acids to produce aluminium salts.

Oxides that can behave as acids or bases depending on their environment are called amphoteric oxides.

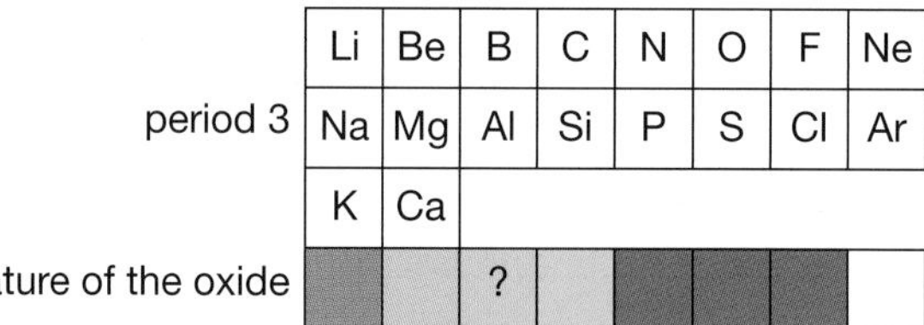

The diagram shows period 3 of the periodic table. On the left-hand side the oxides are very basic; on the right-hand side they are very acidic. Aluminium, in the middle, can be either acidic or basic depending on what it reacts with.

Summary

- Aluminium is relatively unreactive because of a protective oxide layer.
- The protective oxide layer on aluminium can be improved by **anodising**.

8.3	
Co-ordinated	Modular
10.7	22 (15.3)

Titanium

Uses of titanium

Titanium is a **transition metal**. It is strong, **corrosion resistant** and has quite a low **density**. This makes it ideal for the manufacture of aircraft. A titanium alloy was used to make the fuselage of Concorde.

Figure 8.9
Titanium was used in the manufacture of Concorde

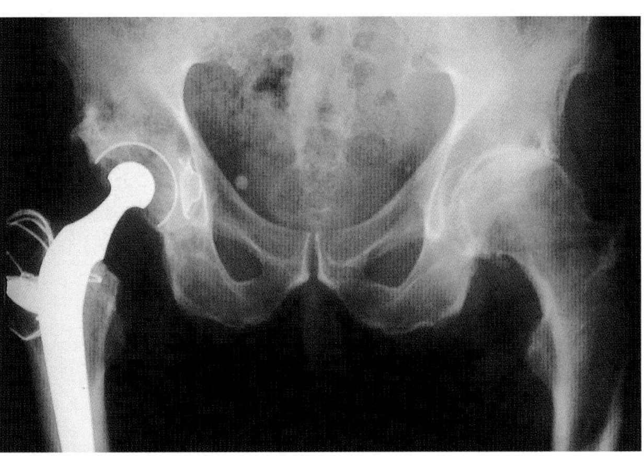

Figure 8.10
This hip joint is made of titanium

Titanium is also used to make replacement hip joints and in various parts of nuclear reactors.

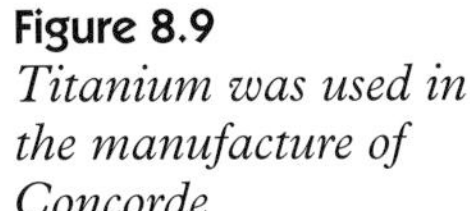

Did you know?

Titanium is the ninth most abundant element in the Earth's crust. It is about 10 times more abundant than copper.

Manufacture of titanium

Titanium is too high in the **reactivity series** to be prepared by **smelting** like iron.

Most reactive metals – like aluminium – are produced by **electrolysis**. Titanium cannot be manufactured this way because its ore is covalently bonded and does not conduct electricity. The common ore of titanium is **rutile** (titanium oxide). The oxide is converted into titanium(IV) chloride ($TiCl_4$) which is also covalently bonded. The titanium(IV) chloride is then reduced by heating it with a more reactive metal like sodium or magnesium.

$$4Na + TiCl_4 \rightarrow Ti + 4NaCl \quad \text{or} \quad 2Mg + TiCl_4 \rightarrow Ti + 2MgCl_2$$

The reaction, which is exothermic, is carried out in special, sealed furnaces (see Figure 2.11). The noble gas argon is put into the furnace to remove all traces of air. If air were present, the sodium or magnesium would react with the oxygen.

Figure 8.11
Titanium is produced in special sealed furnaces

Topic questions

1 Aluminium is above iron in the reactivity series. Explain why aluminium does not corrode as rapidly as iron.

2 What is the name of the process used to improve the protective layer on aluminium?

3 Explain why titanium cannot be produced by:
a) smelting b) electrolysis.

4 Why is argon used in the manufacture of titanium?

Summary

- Titanium is a low density, **corrosion resistant, transition metal.**
- Titanium is used to manufacture some aircraft. It is also used to make artificial hip joints.
- Titanium is produced by heating titanium(IV) chloride with a metal high in the **reactivity series** like sodium or magnesium.

8.4

Co-ordinated	Modular
10.7	22 (15.2)

Steel

The manufacture of pure iron and steel

Cast iron produced by a **blast furnace** is very impure. Among other substances it contains up to 4% carbon. The carbon makes the cast iron very brittle. For many applications cast iron is not suitable. For example, it would not be very sensible to make a hammer from a brittle material like cast iron!

Did you know?

The world's first cast iron bridge was erected at Ironbridge, Shropshire in 1779. It used 400 tonnes of cast iron. One hundred and ten years later, the Eiffel Tower was built in Paris. It used about 7000 tonnes of wrought iron. Wrought iron is better than cast iron but a suitable method of removing the carbon from the iron on a large scale was not available until Sir Henry Bessemer introduced a process in 1859.

To make pure iron, the carbon (and other impurities) have to be removed. This is done by pouring the molten iron from the blast furnace into a special furnace called a converter (Figures 8.12 and 8.13). Scrap iron is put in the converter first to reduce the damage to the furnace. If the scrap iron was not present, the thermal shock of the hot iron hitting the much cooler converter lining could crack the lining. (Rather like pouring boiling water into a milk bottle can cause the glass to break.)

Most of the impurities in iron are **non-metals**. In the furnace these are converted into non-metal oxides by passing pure oxygen through the molten iron.

These are examples of redox reactions.

Carbon is oxidised to carbon dioxide and comes off as a gas. Other non-metals (mainly silicon) also form oxides. These oxides are acidic and are removed by **neutralising** them with limestone (calcium carbonate). The product of this neutralisation is **slag** (mainly calcium silicate) which floats on top of the molten iron and can be removed.

Figure 8.12
Iron is converted to steel in an oxygen furnace

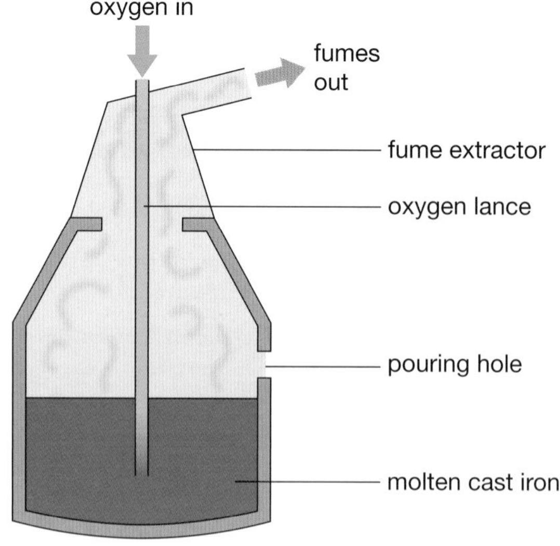

Figure 8.13
The inside of an oxygen converter where cast iron is purified

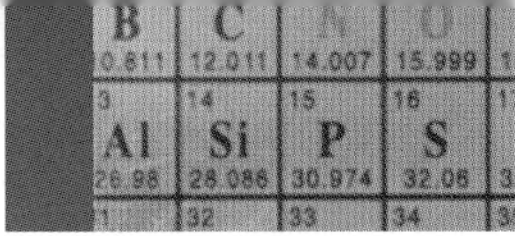

Pure iron is useful in some applications but most of the iron is made into **steel**. To do this, other elements have to be added to the iron. This can either be done by adding the element to the molten iron or adding scrap iron containing known amounts of other elements to the molten iron in the converter.

This process can produce between 400 and 500 tonnes of steel an hour.

Types of steel and their uses

Steel is an **alloy** of iron and carbon. There are many different sorts of steel. The properties of each type of steel depend on how much carbon is added (see Figure 8.14). High carbon steels are very hard and strong but are brittle; they are used where hard wear is required – for example ball bearings and roller bearings. Medium carbon **mild steel** can easily be pressed into shape so is used to make car body panels. **Wrought iron** is almost pure iron and is easily bent into shape so is used for decorative purposes like making iron gates (see Figure 8.15).

Figure 8.14 *Alloys of iron and carbon*

Carbon percentage	Name	A typical use
4–5	cast iron	car engine blocks
0.8–1.5	high carbon steel	ball bearings
0.3–0.8	medium carbon steel	railway lines
<0.3	mild steel	car bodies
very low	wrought iron	ornamental gates

Figure 8.15 *These objects are all made from different types of steel. a) These automotive ball pins are made from high carbon steel. b) Car doors are made from mild steel. c) Ornate gates are made from wrought iron*

The addition of other alloying elements also affects the properties of the steel (see Figure 8.16).

Figure 8.16 *Steels containing other alloying metals*

Alloying metal(s)	Name	A typical use
chromium (up to 5%)	chromium steel	ball bearings
cobalt (up to 10%)	cobalt steel	magnets
chromium (18%) nickel (8%)	stainless steel	cutlery
tungsten (18%) chromium (4%) vanadium (1%)	tool steel	cutting tools for metal working lathes

Did you know?

To make sure the steel has the correct composition, it has to be analysed. This can be done by making an electric arc between a sample of the steel and another electrode. The colour of the spark is used to find the composition of the steel (rather like a computerised flame test – see Chapter 10). The entire process from taking a sample of the molten steel to getting a complete analysis only takes a few minutes.

Electroplating

Like most metals, iron and steel can be **electroplated**. In this process the steel object is completely covered in a thin layer of another metal. The steel object must be free of grease and corrosion. If the plating electrolyte and conditions are correctly controlled, the layer will stick firmly to the steel.

The objects to be electroplated are supported on a suitable jig. The jig and objects are put in the plating **electrolyte** and connected to the negative side of a power supply. The anodes are connected to the positive side of the supply. The anodes are made of the pure metal that is to be plated onto the objects. The electrolyte contains ions of that metal (see Figures 8.17 and 8.18)

The electrode reactions are:

At the anode:

$$M(s) \rightarrow M^{n+}(aq) + ne^-$$

For example with silver plating:

$$Ag(s) \rightarrow Ag^+(aq) + e^-$$

At the **cathode**:

$$M^{n+}(aq) + ne^- \rightarrow M(s)$$

For example with silver plating:

$$Ag^+(aq) + e^- \rightarrow Ag(s)$$

Figure 8.17
An electroplating bath in operation

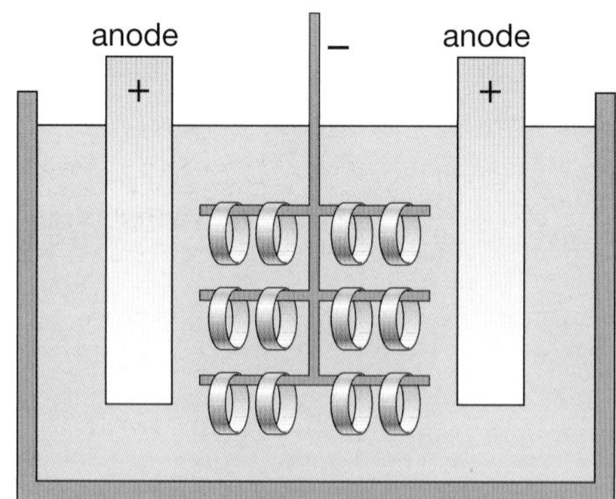

Figure 8.18
Electroplating

Figure 8.19
These items of cutlery have all been electroplated

For every atom of metal that changes into an ion at the anode, one ion changes back to the metal at the cathode. This means that the concentration of the electrolyte doesn't change. All that happens is the anodes gradually 'dissolve' in the **electrolyte**.

Electroplating is used either to protect steel from corrosion or to produce a decorative finish to a product. Steel objects electroplated with silver look like silver objects but are very much cheaper.

Almost any metal can be plated on to the steel object. Silver, gold, nickel and chromium are frequently used.

Electroplating is an example of a **redox reaction**.

At the anode, the metal is oxidised to a positive ion (loss of electrons) and at the cathode, the ions are reduced back to the metal (gain of electrons).

The anode is positively charged because it is deficient in electrons. It works like an oxidising agent by accepting electrons from the metal. The negatively-charged cathode has an excess of electrons. It behaves like a reducing agent by providing electrons to the metal ions in the electrolyte.

Did you know?

Because electroplating is done in aqueous solutions, only metals below hydrogen in the reactivity series can be plated onto steel. This is not a problem as electroplating is usually used to reduce the risk of corrosion – so reactive metals would not be suitable.

If a steel object is placed in a bath of silver electrolyte, a chemical reaction takes place before the electric current can be switched on:

$$Fe(s) + 2Ag^{+}(aq) \rightarrow 2\,Ag(s) + Fe^{2+}(aq)$$

This happens because iron is more reactive than silver. Although only a very small amount of silver is chemically deposited on the steel, it weakens the adhesion between the electroplated silver and the steel. This means the plating is more likely to flake off.

For the highest quality work this is overcome by using several electroplating baths. The first bath would be a very dilute nickel electrolyte to 'flash' plate with a very thin layer of nickel. Because nickel is only slightly lower than iron in the reactivity series, there will be no chemical deposition. The process is repeated with a dilute copper electrolyte then a dilute silver electrolyte to 'flash' plate with each of these metals. Finally the object can be plated in a correct silver plating electrolyte.

Topic questions

1. What is the main difference between cast iron, wrought iron and steel?
2. In the basic oxygen converter to make steel, oxygen gas and calcium carbonate are added to the molten steel. What do each of these substances do?
3. What effect does a high chromium content have on the behaviour of steel?
4. Explain why the concentration of silver ions in a silver plating electrolyte does not change during the silver plating process.

Summary

- **Steel** is an **alloy** of iron and carbon.
- Steel is made from **cast iron** by the basic oxygen process.
- Alloying elements like tungsten and chromium can alter the properties of steel, making it harder or more corrosion resistant.
- **Electroplating** is a process in which metal objects are covered by a thin layer of another metal. The process can be used to protect steel against corrosion.
- Electroplating is a **redox** reaction.

Examination questions

1 The manufacture of sulphuric acid, H_2SO_4, by the Contact Process is shown below.

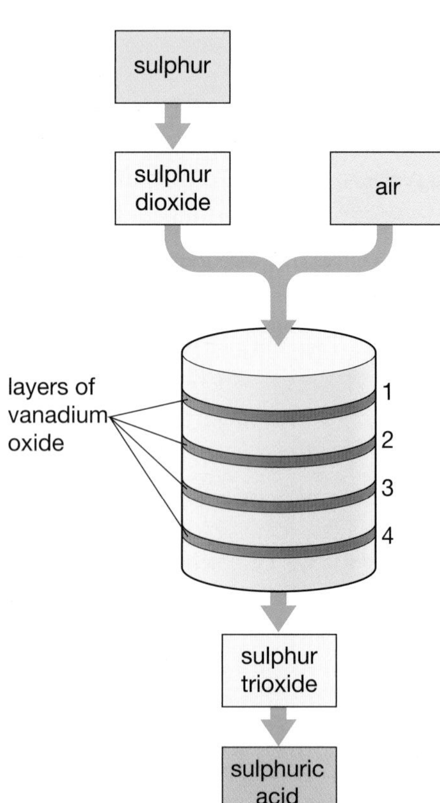

Vanadium oxide layer	Temperature of gas		% SO_2 changed into SO_3
	Before reaction	After reaction	
1	430	590	65
2	440	510	84
3	440	480	92
4	415	440	99

a) i) Name **one** source of sulphur. *(1 mark)*
ii) How is sulphur dioxide made from sulphur? *(2 marks)*

b) i) What is the function of the vanadium oxide? *(1 mark)*
ii) Use the information in the table to suggest why four layers of vanadium oxide are used. *(2 marks)*
iii) What do the differences in temperature before and after each layer show about the reaction to make sulphur trioxide? *(1 mark)*
iv) Suggest what should happen to the 1% of sulphur dioxide that is **not** changed into sulphur trioxide. *(1 mark)*

c) How is sulphuric acid made from sulphur trioxide? *(2 marks)*

d) Work out the mass of sulphuric acid, H_2SO_4, that can be made from 4 tonnes of sulphur. Relative atomic masses H 1; O 16; S 32. *(2 marks)*

2 a) Sulphuric acid is produced in the United Kingdom from sulphur. The three main reactions for the production of sulphuric acid are represented by the equations below.

$$S + O_2 \rightarrow SO_2$$
$$2SO_2 + O_2 \rightarrow 2SO_3$$
$$SO_3 + H_2O \rightarrow H_2SO_4$$

Name **two** raw materials, other than sulphur, needed to make sulphuric acid. *(2 marks)*

b) Blue copper sulphate crystals can be used to show that sulphuric acid is concentrated.

i) What colour change would you **see** when copper sulphate crystals are added to concentrated sulphuric acid? *(1 mark)*

ii) Why does the colour of the crystals change? *(1 mark)*

c) A student diluted some concentrated sulphuric acid with water. The student thought the dilute acid was weak. The teacher said it was still a strong acid.
Why is the acid described as strong? *(1 mark)*

d) The teacher gave the student two solutions. One was a strong acid and the other a weak acid. The solutions were of the same concentration.
Describe a test the student could do to show which solution was the strong acid and which was the weak acid. Give the results of the test with both solutions. *(3 marks)*

3 a) Describe how steel is manufactured using molten iron obtained from the blast furnace. Your answer should include:

- the types of reaction occurring;
- the details of the conditions used;
- energy changes involved. *(5 marks)*

b) Suggest **two** factors which influence the location of plants associated with the manufacture of steel. *(2 marks)*

c) Give **two** reasons why it is important to recycle steel. *(2 marks)*

4 a) The table gives some information about steels. Complete the table by choosing properties from the list.

- soft and easily shaped
- strong but brittle
- resistant to corrosion
- rusts easily

Type of steel	Percentage (%) of carbon	Other elements present	Properties	
			strength	corrosion
high carbon	0.5–1.4	X	....	rusts quite easily
low carbon	0.04–1.15	X	....	....
stainless	0.05–1.10	Cr, Ni	strong and hard	....

(2 marks)

b) Steels are made from molten iron in a furnace. The diagram shows the substances which are added to the molten iron during steelmaking.

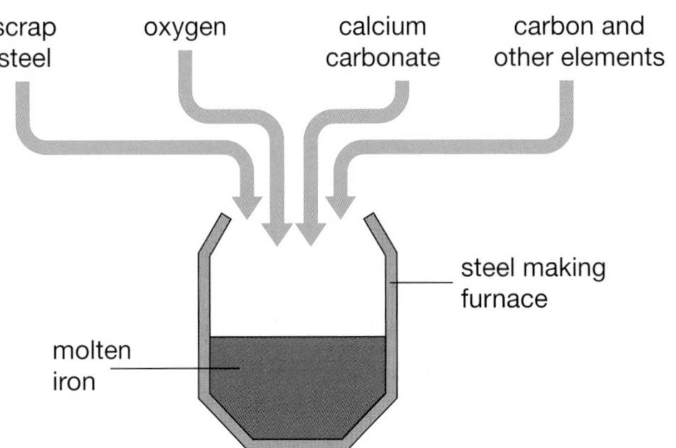

Give a reason why each of the substances is added to the molten iron. *(4 marks)*

c) i) Car bodies made of steel are painted. Explain how this prevents rusting. *(2 marks)*

ii) The steel may be coated with zinc before it is painted. This gives further protection from rusting. Explain how this method of rust prevention works. *(2 marks)*

Chapter 9
Aqueous chemistry

Key terms acid • alkali • anhydrous • anion • artificial fertilisers • base • burette • covalent compounds • electron • hard water • hydrated proton • indicator • ion exchange • ionic compounds • limescale • molar • mole • neutralisation • nucleus • permanent hardness • pH scale • pipette • precipitation • proton • proton acceptors • proton donors • reactivity series • relative atomic mass • relative formula mass • respiration • salt • saturated solution • scum • soap • soft water • solubility • soluble • solute • solvent • strong acid • strong base • temporary hardness • titration • water cycle • weak acid • weak base

9.1 Water is essential to life

Co-ordinated	Modular
10.15	21 (14.1)

Water is the most abundant substance on the Earth's surface. In fact nearly 71% of the Earth's surface is water.

Living things cannot survive without water. About 70% of body tissue is water. In some plant materials the water content can be over 90%. Each person needs to drink about 1.5 litres of water each day to replace what is lost in urine, faeces and sweat. In very hot weather this might go up to as much as 5 litres per day.

Did you know?

The total amount of water on the Earth's surface is about 1 370 000 000 cubic kilometres (1.37×10^9 km^3). That's enough water to fill a garden hosepipe 1.4 million light years long.

Purifying drinking water

Drinking water has to be treated to make it safe. Solids are removed in large filter beds then chlorine is added to kill any bacteria present. Other treatments may also be used depending on the purity of the water source (see Figure 9.1).

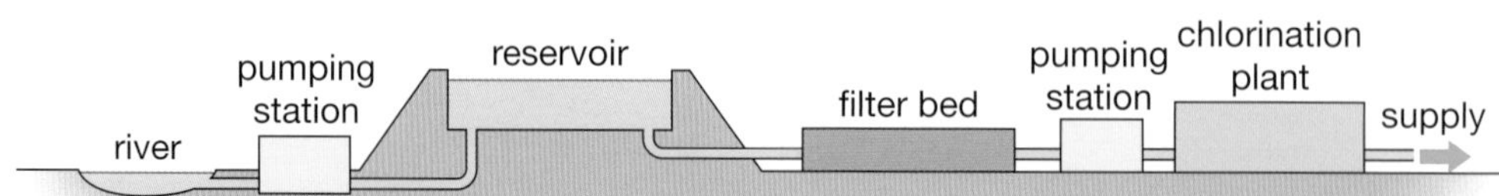

Figure 9.1
(a) The stages in a water treatment plant

Figure 9.1
(b) This water treatment plant is in British Columbia in Canada

This purification process does not remove all impurities. For example nitrates are not removed. Ammonium and nitrate ions can get into the water from the use of **artificial fertilisers**. Rain will wash the ammonium and nitrate ions into the ground. Normally the roots of the plant would remove these ions and use them for growth, but if fertilisers are over used (or if rainfall is very heavy) some of the ions can be washed past the root systems before the roots can absorb them. These ions then drain into streams and can get into drinking water. The water companies carefully monitor the nitrate content of drinking water as too much is harmful, especially to bottle-fed babies.

Did you know?

Nitrates in water are turned into nitrites in the intestine. They can re-enter the body and react to form carcinogenic (cancer-forming) compounds called nitrosamines.

In babies the nitrites can interfere with haemoglobin, limiting its ability to carry oxygen. The baby develops a fatal form of anaemia called 'blue-baby syndrome'. There have been no deaths from this recorded in the UK since 1948.

It is possible to remove nitrates from domestic drinking water by ion exchange (see section 9.3) or reverse osmosis equipment, but these are fairly expensive. Increasing numbers of people drink bottled water to reduce their intake of potentially harmful substances in water. But bottled water often has a much higher level of bacteria than tap water.

Summary

- Water is essential to life.

9.2 The water cycle

Co-ordinated	Modular
10.15	21 (14.1)

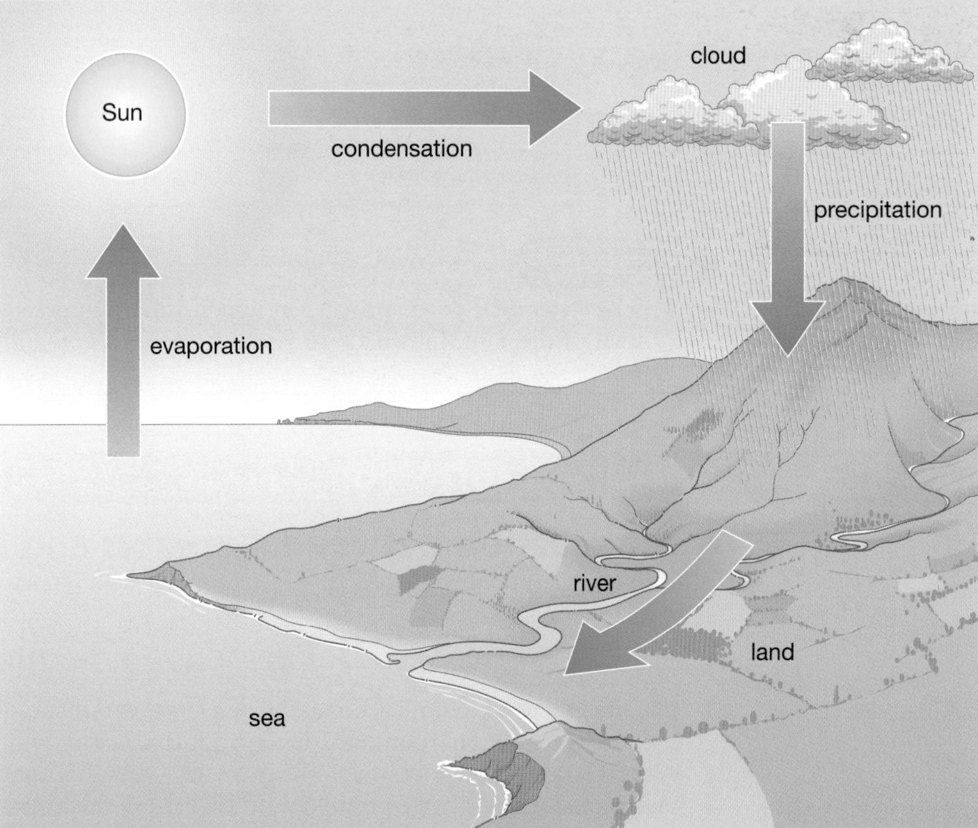

Figure 9.2
The water cycle

Water in rivers, lakes and seas is evaporated by the heat of the Sun. The pure water vapour rises and cools. As it cools, the water vapour condenses to form clouds. As the clouds cool further, rain, and other forms of precipitation like snow and hail, are produced. This process is called the **water cycle**.

The rain that falls starts off as fairly pure water. As it falls it will dissolve some carbon dioxide from the air. Once it gets into the ground it will dissolve a lot more carbon dioxide. It may also dissolve other substances from the soil. The water can also dissolve substances from the rocks as it passes through them.

One of the substances dissolved by the water as it passes through the rock is sodium chloride (common salt). The water will not have much salt in it as it flows down river but over many millions of years, the amount has built up in the sea so that seawater now contains quite a high salt content.

Figure 9.3 *The extraction of sea salt from sea water*

Did you know?

Seawater contains about 3% sodium chloride. It also contains many other substances.

Magnesium	0.13%
Calcium	0.04%
Potassium	0.04%
Bromine	0.006%

There is gold in seawater. The percentage is very low (about 6×10^{-10}%), but the total amount of gold in all the oceans in the world is about 8 million tonnes. It would, however, cost more to get the gold out of the water than it would be worth.

Summary

- The **water cycle** is the process in which water, **evaporated** from the sea by the heat of the Sun, falls back to Earth by **precipitation** and eventually flows back to the sea.

9.3 Hard and soft water

Co-ordinated	Modular
10.15	21 (14.1)

The reaction between rainwater and limestone

Rainwater containing carbon dioxide can react with limestone in the following reaction.

$$CaCO_3(s) + H_2O(l) + CO_2(aq) \rightarrow Ca(HCO_3)_2(aq)$$

The limestone (calcium carbonate, $CaCO_3(s)$) is not soluble in water but it reacts with the water and carbon dioxide to produce calcium hydrogencarbonate ($Ca(HCO_3)_2(aq)$) which *is* soluble in water. Over thousands of years this process has gradually made huge potholes and caverns in the limestone rocks.

Figure 9.4
This cave formed when the limestone dissolved in the acidic rain water

This reaction is reversible. Once the very dilute calcium hydrogencarbonate solution gets into the open cavern, some of the carbon dioxide in the water is given off. This causes the reaction to go in the other direction.

$$Ca(HCO_3)_2(aq) \rightarrow CaCO_3(s) + H_2O(l) + CO_2(g)$$

The limestone forms where the water emerges from cracks in the rock. This causes stalagmites and stalactites to form.

Did you know?

The reason why there is not a large amount of calcium hydrogencarbonate in the oceans is because it decomposes in the warm seas to form calcium carbonate solid which is deposited as layers and eventually forms rock. Sea creatures also remove the calcium to form their shells of calcium carbonate.

The cause of hard and soft water

Water containing dissolved calcium ions (Ca^{2+}) is called **hard water**. If drinking water is extracted from rivers that have their source in limestone areas, the water will contain calcium hydrogencarbonate and will be hard. There is another substance that can make water hard. That substance is calcium sulphate. Calcium sulphate is found in the rocks gypsum and alabaster and is slightly soluble in water. Rocks containing magnesium behave in a similar way and the magnesium ion (Mg^{2+}) can also make water hard.

Hardness caused by calcium hydrogencarbonate is called **temporary hardness** because it can easily be removed by heating the water

$$Ca(HCO_3)_2(aq) \rightarrow CaCO_3(s) + H_2O(l) + CO_2(g)$$

Hardness caused by calcium sulphate cannot be removed by heating and is called **permanent hardness**.

Water containing no calcium or magnesium ions is called **soft water**.

The effects of hard water

a) Good effects

The substances dissolved in hard water are good for your health. Calcium ions are needed to build healthy teeth and bones. There is also evidence that calcium ions help to reduce the incidence of heart disease.

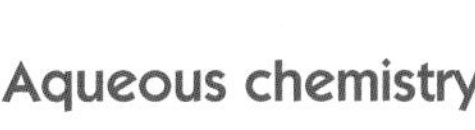

Figure 9.5
The scum around this sink has been formed because of the calcium salts in the water

b) Bad effects

1 **Soaps** are compounds that help to remove oily material from clothes. They are the sodium salts of carboxylic acids (see Chapter 7). In soaps the carboxylic acids contain long chains of carbon atoms. Sodium soaps are soluble in water. If the water contains calcium ions, the following reaction occurs between the soap and the calcium ions.

sodium 'soap' + calcium ions → calcium 'soap' + sodium ions
(soluble) (insoluble 'scum')

The calcium soap is not soluble in water. It forms a greasy '**scum**' that sticks to clothes. The reaction stops the soap from working so a lot of extra soap is required to produce lather. This makes the washing process much more expensive.

2 Hard water can also cause a build up a deposit of '**limescale**' in kettles and boilers. If the hardness is caused by calcium hydrogencarbonate, the scale is a deposit of calcium carbonate formed by the reverse of the reaction that put the hardness in the water.

$$Ca(HCO_3)_2(aq) \rightarrow \underset{\text{limescale}}{CaCO_3(s)} + H_2O(l) + CO_2(aq)$$

The deposit reduces the efficiency of the kettle or boiler because it is not a good conductor of heat. This increases the cost of producing hot water. Scale can also accumulate in boilers and may eventually damage the boiler.

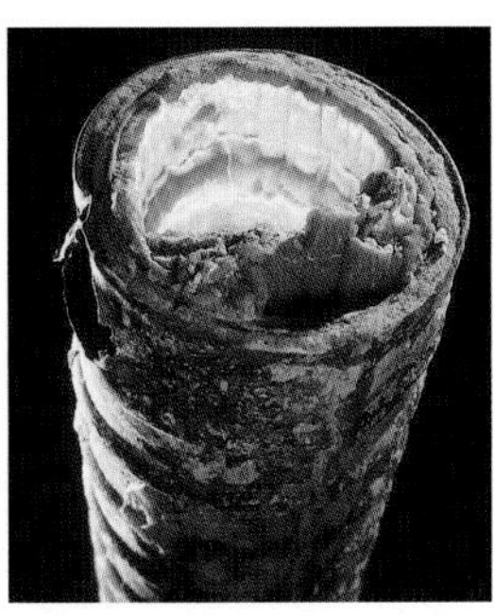

Figure 9.6
You can see how limescale has built up in this domestic metal pipe

Removing hardness

To remove hardness, the dissolved calcium and magnesium ions have to be removed. There are several ways of doing this.

1 Precipitating the calcium and magnesium ions

The addition of sodium carbonate will remove the calcium and magnesium ions. As the sodium carbonate dissolves in the hard water, the carbonate ions react with the calcium ions to produce calcium carbonate which is not soluble in water. Calcium carbonate forms as a precipitate.

$$Na_2CO_3(aq) + Ca(HCO_3)_2(aq) \rightarrow 2NaHCO_3(aq) + CaCO_3(s)$$

Washing 'soda' and bath salts contain sodium carbonate and this is how they work to soften water.

2 Using ion exchange to remove calcium and magnesium ions

Ion exchange columns contain a porous material that has a lot of negatively-charged sites on it. Attached to these sites are positive ions, usually sodium ions (Na^+) or hydrogen ions (H^+). When water containing calcium or magnesium ions passes down the column, these ions displace the sodium or hydrogen ions. The material holds the calcium or magnesium ions so the water that comes from the bottom of the tube has no calcium or magnesium ions in it. This is now soft water.

After a while the material in the column has to be renewed. This is done by reversing the process and washing out the calcium ions with a concentrated solution of sodium ions (using salt solution) or hydrogen ions (using an acid like hydrochloric acid).

Did you know?

Soapless detergents are not affected by hard water. This is because both the sodium and the calcium form of the detergent are soluble in water so no scum forms and the detergent works normally.

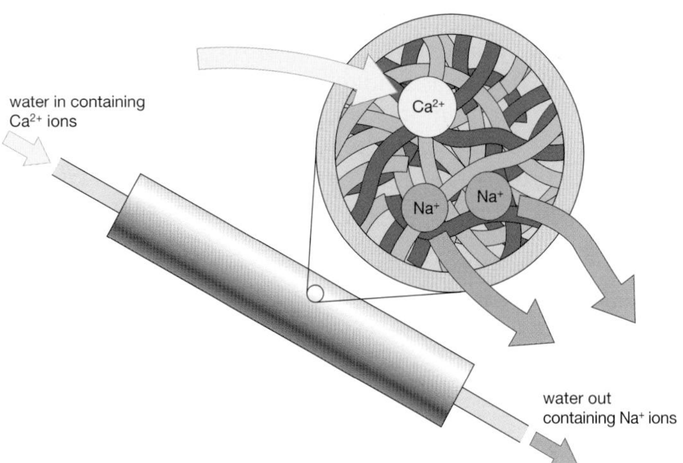

Figure 9.7
What happens in an ion exchange column

Topic questions

1 There are two main processes used in the purification of drinking water. How does each of these processes make the water pure?

2 Explain how nitrates can get into drinking water.

3 Outline the water cycle using the terms *evaporation*, *condensation* and *precipitation*.

4 Which **two** ions are mainly responsible for hard water?

5 a) Which substance in hard water can be removed by heating the water?
b) What name is given to hardness that can be removed by heating water?

6 Explain how hardness can be removed by:
a) washing 'soda' b) ion exchange.

Summary

- **Hard water** contains calcium and/or magnesium ions.
- Hard water causes '**scum**' with soap and harmful '**limescale**' in kettles and boilers.
- Soft water contains few or no calcium and/or magnesium ions.
- Hard water can be **softened** by removing the calcium and/or magnesium ions

9.4	
Co-ordinated	Modular
10.15	21 (14.2)

Solubility

Water is a powerful **solvent**. Many substances are **soluble** in water. The ability of water to dissolve substances is widely used in industry. For example, in the metal plating industry (see Chapter 8) water is the solvent used for the electrolyte. In industry cost is important and water is a very cheap solvent. But not only is it cheap, for many applications water is also the best solvent. Water also has a high thermal capacity. This means it is very effective as a coolant. Water is still used as a coolant in many industrial processes (e.g. power stations) and in the engines of many motor vehicles. It is the coolant properties of water that enable it to be used effectively to put out fires.

The solubility of gases

Many gases are soluble in water. But water will not dissolve an infinite amount of gas – there is a maximum. The maximum amount of gas that will dissolve in water is called the **solubility**.

Figure 9.8 shows the solubilities of some gases in water. The figures show the maximum number of millilitres of gas that will dissolve in 1 millilitre of water. It is clear that there is a huge range in solubility. Nitrogen and oxygen are only slightly soluble but ammonia is very soluble. The table also shows that gases are more soluble in cold water than in hot water. This is why a glass of cold water left on a bedside table usually has small gas bubbles in it in the morning. Overnight the water warms up and the gases dissolved in it become less soluble (see Figure 9.9).

Figure 9.8
The solubility of some gases in water

Gas	Solubility in water (ml/ml)	
	0°C	60°C
nitrogen	0.024	0.01
oxygen	0.049	0.019
carbon dioxide	1.71	0.36
chlorine	4.61	1.01
hydrogen chloride	507	339
ammonia	1299	about 230

Figure 9.9
Bubbles have developed in the glass of water overnight

Figure 9.10
The bubbles rise when a bottle of fizzy drink is opened

Gases are also more soluble in water if the pressure is increased. This is why fizzy drinks always bubble when the top is opened (see Figure 9.10).

Carbon dioxide

Fizzy drinks are made using carbonated water – that is water with carbon dioxide dissolved in it under pressure. Carbon dioxide is used because it is readily available, cheap and non-toxic. It is better than oxygen or nitrogen because it is about 30 times more soluble and so makes the drinks fizzier.

Did you know?

Sparkling wines like champagne are also fizzy because of dissolved carbon dioxide. But in wines, the gas is not added, it is produced naturally in the wine during the fermentation process.

Oxygen

Although oxygen is not very soluble in water, the amount that does dissolve is important to aquatic life. Fish and other animals extract the oxygen they need for **respiration** directly from the water. If the water is too hot then they will die because less oxygen is available in the water. This is why power stations have to be very careful to ensure that the cooling water they discharge into rivers and lakes is not too hot. The huge cooling towers at power stations are for this purpose.

Figure 9.11
The cooling towers at a power station

Figure 9.12
Household bleach contains chlorine water

Did you know?

If oxygen was more soluble in water then the oceans could dissolve more and there would be less oxygen in the atmosphere. At the average temperature of the oceans, 1 millilitre of water will dissolve 0.04 millilitres of oxygen. If the solubility of oxygen were doubled then the oxygen content of the atmosphere would drop from 21% to about 14%.

Chlorine

Chlorine is fairly soluble in water (see Figure 9.8). A solution of chlorine in water is called 'chlorine water'. It is used to bleach materials and to kill bacteria. Most domestic bleaches and some toilet cleaners contain chlorine water. Chlorine is also dissolved in the water of public swimming pools to reduce the risk of diseases being spread.

Figure 9.13
A chlorination plant at a swimming pool

The solubility of solids

Most **ionic compounds** are soluble in water, though some, for example calcium carbonate, are not. **Covalent compounds** are usually not soluble in water, though here too, there are exceptions, for example sugar.

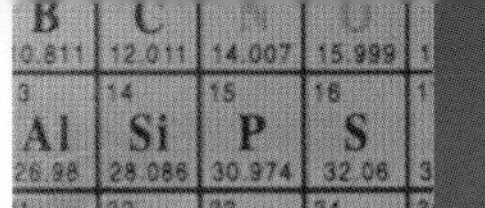

Aqueous chemistry

When a solid dissolves in water it forms a solution. Water is the **solvent** and the solid is the **solute**. The same principle applies to solids as to gases – there is a maximum amount of solute that will dissolve in a solvent. This maximum quantity is called the solubility. It is measured as the number of grams of solute that will dissolve in 100 grams of solvent. The solubility of solids depends on the temperature. Unlike gases, solids are usually more soluble in hot solvents. So when quoting the solubility of a substance it always necessary to specify the temperature.

A solution that has got the maximum possible amount of solute in it is called a **saturated solution**. If a saturated solution is cooled, less of the solute will be able to dissolve. The excess solute will separate from the solution as a solid – often in the form of crystals.

Figure 9.14
Table of the solubility of some solids at different temperatures

Substance	Temperature/°C						
	0	10	20	40	60	80	100
potassium nitrate	13.3	20.9	31.6	63.9	110	169	246
sodium chloride	35.7	35.8	36	36.6	37.3	38.4	39.8
potassium chloride	28.1	31.2	34.2	40	45.8	51.3	56.3
calcium hydroxide	0.185	0.176	0.165	0.141	0.116	0.094	0.077
copper(II) sulphate	14.3	17.4	20.7	28.5	40	55	75.4

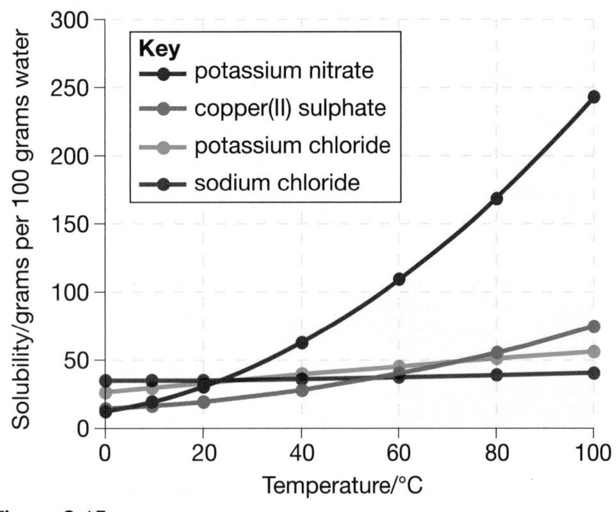

Figure 9.15
Solubilities of some common substances

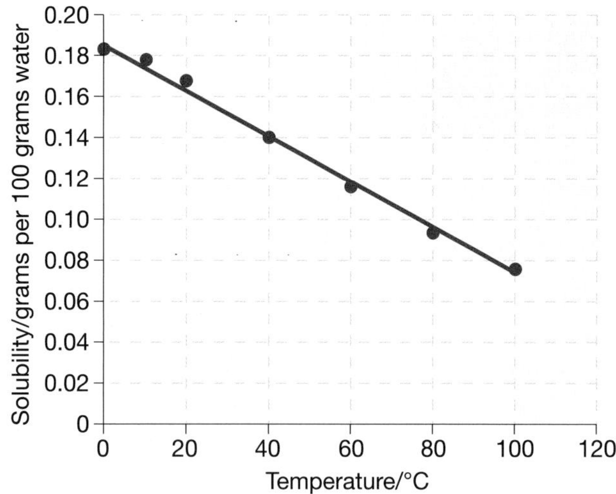

Figure 9.16
Solubility of calcium hydroxide

With solubilities there is a wide range of behaviour patterns. Some have quite low solubility at 0°C but very high solubility at 100°C (for example potassium nitrate), for others the solubility does not change much with the temperature (for example sodium chloride) and some have very low solubility (for example calcium hydroxide.) Calcium hydroxide shows another unusual pattern. Its solubility decreases as the temperature goes up.

Did you know?

Barium compounds are very poisonous, but doctors get people to drink 'barium meals' so they can have their stomach X-rayed. Barium meals contain a lot of barium sulphate. It does not poison the person who drinks it because barium sulphate is not very soluble in water. Only 0.00025 grams of barium sulphate will dissolve in 100 grams of water – this is not enough to be toxic.

Using a solubility curve

Example 1

A saturated solution of potassium nitrate is made by dissolving the solute in 100 grams of water at 90°C. The solution is then cooled to room temperature (20°C). What mass of potassium nitrate crystals would be produced?

From the graph of the solubility of potassium nitrate:

Amount of potassium nitrate soluble in 100 g water at 90°C	205 g
Amount of potassium nitrate soluble in 100 g water at 20°C	32 g

When a saturated solution of potassium nitrate at 90°C is cooled to 20°C, then 205 – 32 = **173 g** of solid potassium nitrate are produced.

Example 2

A solution of copper(II) sulphate solution is boiled to remove some of the water. When the volume has been reduced to 500 ml, the solution is left to cool. At 80°C crystals of copper(II) sulphate begin to form. What mass of copper(II) sulphate will be formed when the solution has cooled to 20°C?

The solution becomes saturated at 80°C (when the crystals first begin to form).

From the graph, at 80°C 100 g of water will dissolve 55.0 grams of copper(II) sulphate.

100 g of water have a volume of about 100 ml, so 500 ml of copper sulphate solution will contain about 500 g of water.

If 100 g of water contain 55.0 g of copper(II) sulphate, then 500 g will contain about $5 \times 55 = 275$ g of copper(II) sulphate.

At 20°C 100 g of water will dissolve about 21 g of copper(II) sulphate.

So 500 g of water will contain about $5 \times 21 = 105$ g of copper(II) sulphate.

At 80°C there were 275 g of copper(II) sulphate dissolved in the water and at 20°C there were 105 g still dissolved in the water.

This means that the mass of the crystals produced must be 275 – 105 = **170 grams.**

Topic questions

1 What is meant by the words:

a) solute?
b) solvent?
c) solution?

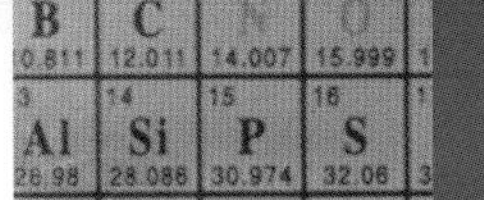

2 Complete the following sentences using the words in the box.

less	more

Most gases are _______ soluble in hot water than in cold water.

Solids are usually _______ soluble in hot water.

3 What is the definition of solubility?

4 The table shows the solubility of ammonium chloride at different temperatures.

a) Draw a graph of the solubility of ammonium chloride against temperature.
b) What is the solubility of ammonium chloride at 25°C?
c) What is the solubility of ammonium chloride at 75°C?
d) How much ammonium chloride will crystallise from 200 ml of solution that is saturated at 75°C when it is cooled to room temperature of 25°C?

Temp /°C	Solubility /g per 100 g
0	29.4
10	33.3
20	37.2
40	45.8
60	55.2
80	65.6
100	77.3

5 Air in the atmosphere contains about 20% oxygen. Explain why air dissolved in water contains about 33% oxygen.

Summary

- The **solubility** of a substance at a particular temperature is the maximum number of grams of solute that will dissolve in 100 grams of solvent at that temperature.
- Gases are less soluble in hot water than in cold water.
- Solids are usually more soluble in hot water than in cold water.

9.5	
Co-ordinated	Modular
10.15	21 (14.3)

Acids and bases

What is an acid?

The Arrhenius idea of an acid

Arrhenius was a Swedish scientist. In 1887 he put forward his theory of ionisation. The theory suggested that many substances in solution were dissociated into ions. His idea was not popular. Other scientists argued that it was not possible for a substance to break down in this way because of the amount of energy it required. For example when sodium reacts with chlorine to produce sodium chloride, a lot of energy is released. The scientists thought it was impossible that water could provide enough energy to reverse this change. Arrhenius pointed out that sodium ions and chloride ions were not the same as atoms and that the amount of energy required to separate ions was much less.

Figure 9.17
Arrhenius

Arrhenius extended his idea to acids and bases. He defined an **acid** as 'a substance which on dissolving in water dissociates to produce hydrogen ions'.

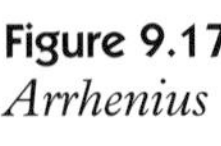

Did you know?

Most non-metal oxides are acidic even though the non-metal itself is not acidic. This is why early scientists thought that acidity was caused by oxygen. In fact the name 'oxygen' means 'acid creator'.

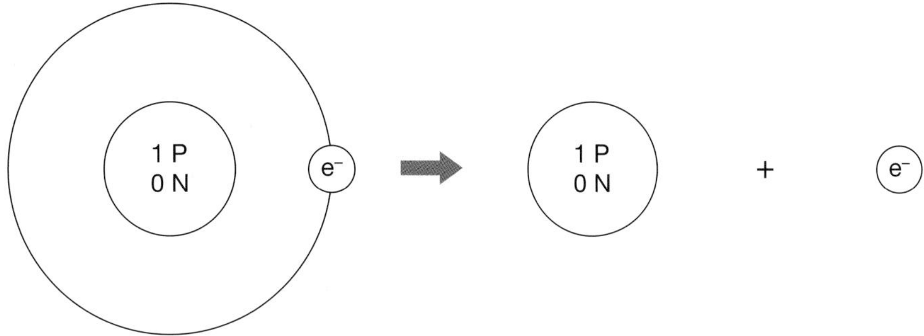

Figure 9.18
How a hydrogen atom becomes a hydrogen ion

Figure 9.18 shows how a hydrogen atom can separate into a hydrogen ion and an **electron**. The hydrogen ion is just the **nucleus** of the hydrogen atom and contains only one proton. So a hydrogen ion is just a **proton**.

Hydrogen chloride is a gas. It is very soluble in water (see Figure 9.8). When it dissolves in water it forms a very acidic solution called hydrochloric acid. Figure 9.19 shows the stages in this process.

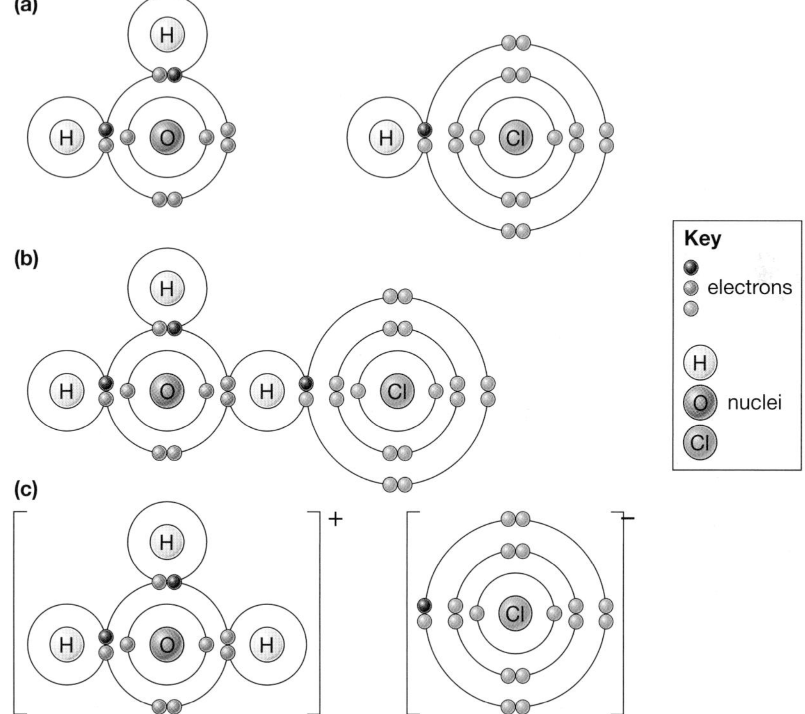

Figure 9.19
How hydrogen chloride reacts with water to form hydrochloric acid

Figure 9.19 a) shows the covalent structures of hydrogen chloride and water. Because hydrogen chloride is covalently bonded, there are no hydrogen ions present so it is not acidic. When hydrogen chloride dissolves in water, a reaction takes place. In this reaction the hydrogen atom on the hydrogen chloride molecule attaches to one of the unused pairs of electrons (called 'lone pairs') on the oxygen atom in the water molecule. This breaks the bond between the hydrogen atom and the chlorine atom, as shown in Figure 9.19 c). The products are the H_3O^+ ion and the Cl^- ion. The H_3O^+ ion is just a **hydrated proton** (hydrogen ion) and is more usually written as $H^+(aq)$.

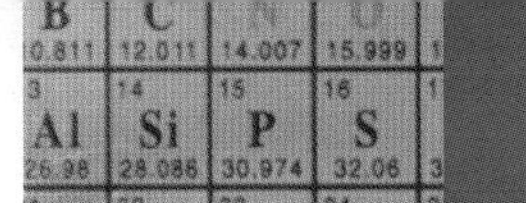

Hydrogen chloride is, therefore, not an acid until it dissolves in water.

Figure 9.20
(a) Blue litmus paper on concentrated sulphuric acid. (b) What happens when one drop of water is added to the litmus paper

The same is true of sulphuric acid. Concentrated sulphuric acid is not acidic. If a piece of blue litmus paper is placed on the surface of concentrated sulphuric acid, it remains blue (see Figure 9.20a). There are no hydrogen ions in concentrated sulphuric acid. But if one drop of water is added, the litmus paper turns red instantly (Figure 9.20b). When water is present the substance becomes acidic.

As a general rule, water must be present for a substance to act as an acid.

All the reactions that are typical of acids depend upon the acid having hydrated protons available (hydrogen ions). For this reason acids can be defined as **proton donors**.

What is a base?

The term '**base**' was first used in the 1770s by Rouelle to describe substances that react with acids to form salts. The familiar relationship is:

$$\text{base} + \text{acid} \rightarrow \text{salt} + \text{water}$$

Using this definition, a base is either a metal oxide or a metal hydroxide. The reaction between a base and an acid is called **neutralisation** because the 'power' of the acid is neutralised by the base. The key reaction that occurs is either:

$$2H^+(aq) + O^{2-} \rightarrow H_2O \quad \text{or} \quad H^+(aq) + OH^- \rightarrow H_2O$$

In each of these reactions the 'active ingredient' of the base uses a hydrated proton from the acid. So if acids are proton donors then bases can be defined as **proton acceptors**.

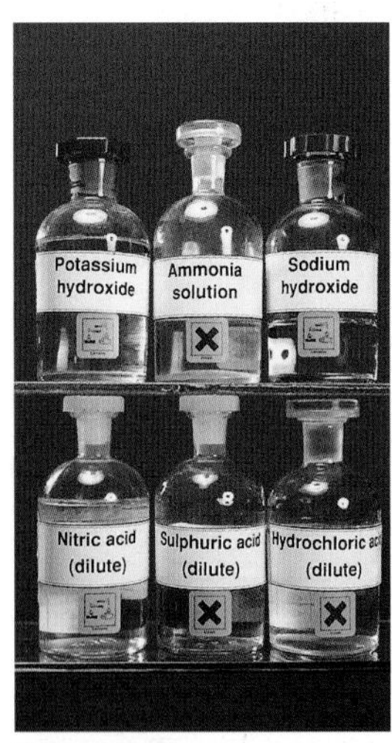

Figure 9.21
Some common acids and bases

Alkalis

Most metal oxides and hydroxides are insoluble in water. Metal oxides that do dissolve in water react with it to form a hydroxide. This is illustrated by the reaction between calcium oxide (quicklime) and water to produce calcium hydroxide (slaked lime).

$$CaO(s) + H_2O(l) \rightarrow Ca(OH)_2(aq)$$

An **alkali** is a soluble base so all alkalis are metal hydroxides.

Arrhenius defined an alkali as 'a substance which on dissolving in water dissociates to produce OH^- ions.'

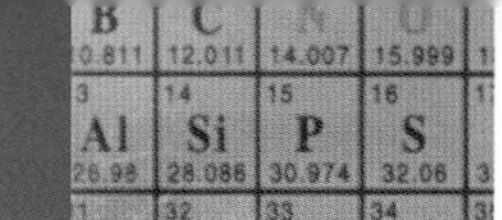

The Lowry and Brønsted theory of acids and bases

Figure 9.22
a) Lowry and
b) Brønsted

The Arrhenius idea of acids and bases is limited to the behaviour of substances when dissolved in water. The Lowry and Brønsted theory proposed in 1923 extends the Arrhenius idea to other situations.

Arrhenius defined an acid as 'a substance which on dissolving in water dissociates to produce hydrogen ions' and an alkali as 'a substance which on dissolving in water dissociates to produce hydroxide ions.' The Lowry and Brønsted theory defines an acid as 'a substance that can give up a proton to a base' and a base as 'any substance which can combine with a proton.' There is no reference to water in this definition.

Looking back at Figure 9.19, the Arrhenius definition of an acid would fit hydrogen chloride but nothing in the diagram would meet his definition of a base. Using the Lowry and Brønsted theory, hydrogen chloride still fits the definition of an acid but this time water fits the definition of a base because it combines with a proton.

The Arrhenius theory took a long time to be accepted because it was an entirely new idea that did not seem to fit with the scientific ideas of the time. But the Lowry and Brønsted theory was readily acceptable because it was a refinement of an existing idea (Arrhenius' idea) and because it also explained behaviours with non-aqueous solvents that the Arrhenius theory did not cover.

Strong and weak acids and bases

Strong and weak acids

Hydrogen chloride dissociates in water to form hydrochloric acid:

$$HCl + H_2O \rightarrow H_3O^+ + Cl^-$$

which can be simplified to:

$$HCl(aq) \rightarrow H^+(aq) + Cl^-(aq)$$

Ethanoic acid dissociates in the same way in the following reaction:

$$HOOC.CH_3(aq) \rightarrow H^+(aq) + {}^-OOC.CH_3(aq)$$

In hydrochloric acid almost all the hydrogen chloride dissociates into ions but in ethanoic acid, only about 0.3% of the acid is dissociated into ions.

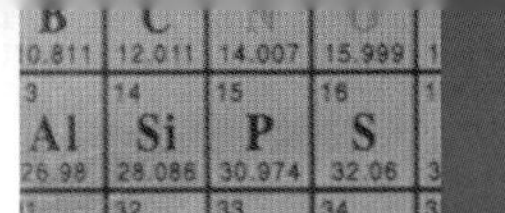

Acids which are highly ionised are called **strong acids** while those which are only slightly ionised are called **weak acids**. By this definition, hydrochloric acid is a strong acid and ethanoic acid is a weak acid. Figure 9.23 summarises this idea.

Figure 9.23
The ionisation of strong and weak acids

$$\mathrm{HA(aq)} \rightleftharpoons \mathrm{H^+(aq) + A^-(aq)}$$
Ionisation of a strong acid

$$\mathrm{HA(aq)} \rightleftharpoons \mathrm{H^+(aq) + A^-(aq)}$$
Ionisation of a weak acid

Hydrochloric, sulphuric and nitric acids are examples of strong acids. Ethanoic, citric and carbonic acids are all examples of weak acids.

Strong and weak bases

Exactly the same idea applies to bases. **Strong bases**, like sodium and potassium hydroxides, are almost completely ionised in water but **weak bases**, like ammonia solution, are only slightly ionised.

The pH scale

A cubic decimetre of **molar** hydrochloric acid contains 1 **mole** of hydrogen chloride. Because the acid is 100% ionised, we can say that it contains 1 mole of hydrogen ions. The concentration of hydrogen ions is, therefore, 1 M (1 molar).

For the same volume of ethanoic acid of the same concentration, the number of moles of hydrogen ions is only about 0.003. In this case the concentration of hydrogen ions is 0.003 M (0.003 molar). In weaker acids or in more dilute solutions, the concentration of hydrogen ions is much lower.

Figure 9.24
The pH values of different acid concentrations

Concentration of hydrogen ions/mol dm^{-3}	pH
1×10^{-7}	7
1×10^{-6}	6
5×10^{-6}	5.3
1×10^{-4}	4
5×10^{-4}	4.3
0.001	3
0.005	2.3
0.01	2
0.05	1.3
0.1	1
0.5	0.3
1	0

The **pH scale** is a convenient way of representing very low concentrations of hydrogen ions. It runs from 0 to 14 and is defined as:

$$\mathrm{pH} = -\log_{10}[\text{concentration of hydrogen ions}]$$

(Don't worry about the meaning of '$-\log_{10}$' – this is easily worked out with a calculator.)

The strongest acids have the lowest pH – around 0–1 on the scale. Strong alkalis have the highest pH – around 13–14. Neutral solutions have a pH of 7.

A molar solution of hydrochloric acid has a hydrogen ion concentration of 1 mol dm^{-3} so it has a pH of 0.

A molar solution of ethanoic acid has a hydrogen ion concentration of about 0.003 mol dm^{-3} so it has a pH of 2.52.

For the same concentration of acid, weak acids will have a higher pH.

Topic questions

1 a) What is the Arrhenius definition of an acid?
b) Use this definition to explain why acids are called proton donors.

2 Explain why hydrogen chloride gas will not turn dry blue litmus paper red but will turn damp blue litmus paper red.

3 Complete the following general equation: acid + base → ______ + ______

4 What is the difference between an alkali and a base?

5 When ammonia gas dissolves in water, a reaction takes place. The equation for the reaction is:

$$NH_3 + H_2O \rightleftharpoons NH_4^+ + OH^-$$

Use the Lowry and Brønsted theory to decide which of these substances are acids and which are bases.

6 What is the difference between a strong acid and a weak acid?

7 a) A solution of a strong base has a pH of 12. What would you expect the pH of a solution of a weak base of the same concentration to be?

A 7
B between 7 and 12
C 12
D greater than 12

b) Explain your answer to part a).

Summary

- **Acids** are substances that dissociate in water to produce hydrogen ions (Arrhenius).
- Substances like hydrogen chloride and concentrated sulphuric acid only behave as acids when they are dissolved in water.
- A hydrogen ion is a proton so acids produce **hydrated protons.**
- Acids are **proton donors** (Lowry and Brønsted).
- **Bases** are substances that **neutralise** an acid to produce a salt and water.
- Bases are **proton acceptors** (Lowry and Brønsted).
- **Alkalis** are soluble bases.
- Alkalis dissociate in water to produce OH^- ions (Arrhenius).
- **Strong acids** and **strong bases** are almost 100% ionised in water.
- **Weak acids** and **weak bases** are only slightly ionised in water.
- The **pH scale** is a way of measuring the concentration of hydrogen ions in a solution.

9.6 Making salts

Co-ordinated	Modular
10.15	21 (14.4)

Reacting a base with an acid

When an acid reacts with a base, the products are a **salt** and water. Because acids only behave as acids when they are dissolved in water, water is always present in the reactants in a neutralisation reaction such as this. We can therefore ignore water when we consider the products of the reaction. In effect the method produces just a salt. To ensure that the salt produced is pure, it is vital that there is no excess acid or base dissolved in the solution.

Using an insoluble base

Because the base is not soluble in water, it is easy to ensure that there is no excess acid or base in the final solution of the product salt. If excess base is added, all the acid will be neutralised. The unreacted base can then be removed by filtration leaving just a solution of the salt.

This method does not work if the salt produced is insoluble in water so it could not be used to make lead(II) sulphate.

Making copper(II) sulphate from copper(II) oxide and dilute sulphuric acid

$$H_2SO_4(aq) + CuO(s) \rightarrow CuSO_4(aq) + H_2O(l)$$

Figure 9.25
Copper(II) sulphate is prepared by dissolving copper oxide in dilute sulphuric acid

The stages involved in this process are summarised below:

a) A beaker containing 50 ml of 2M sulphuric acid is heated and copper(II) oxide is slowly added. After each addition, the mixture is stirred until the copper(II) oxide has dissolved. The solution turns blue and each addition of copper(II) oxide increases the intensity of the blue colour as more copper(II) sulphate is formed. It takes about 8 g of copper(II) oxide to neutralise the acid completely.

b) Once that quantity has been added, all the sulphuric acid has been neutralised. The solution is filtered to remove the unreacted copper(II) oxide.

c) The solution is then boiled to remove some of the water. When the total volume is down to about 20 ml, the beaker is left to cool.

d) Blue crystals of copper(II) sulphate are produced as the solution evaporates.

Using a metal carbonate

This method is essentially the same as the method above. The only obvious difference is that carbon dioxide gas is given off in the process. This method also does not work if the salt produced is insoluble in water so it could not be used to make calcium sulphate.

Making copper(II) chloride from copper(II) carbonate and dilute hydrochloric acid

$$2HCl(aq) + CuCO_3(s) \rightarrow CuCl_2(aq) + H_2O(l) + CO_2(g)$$

Figure 9.26
Copper(II) chloride is prepared by dissolving copper(II) carbonate in dilute hydrochloric acid

Using a soluble base

With a soluble base it is more difficult to make sure that neither reactant is in excess. If the concentrations of the two solutions are accurately known, the correct volumes can be calculated. The volume of acid and alkali that neutralise each other can be measure by a method called **titration**.

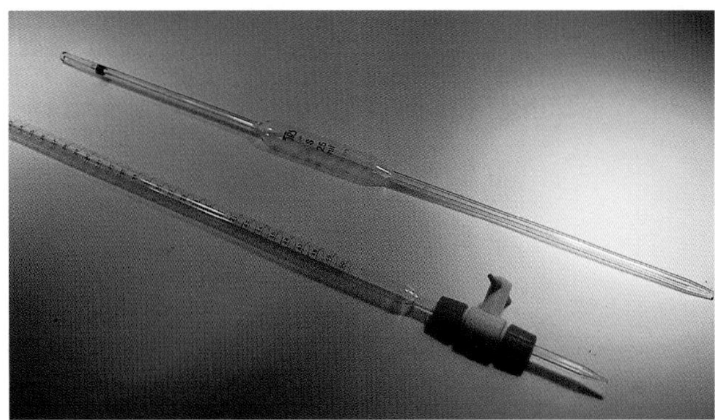

Figure 9.27
A pipette and a burette are used when carrying out a titration

In a titration, a volume of alkali, measured accurately with a **pipette**, is added to a conical flask and a few drops of a suitable **indicator** are added. (Litmus is not an ideal indicator for this process. It is better to use methyl orange or phenolphthalein.) Acid from a **burette** is now added carefully until the indicator changes colour. The volume of acid used is measured. The same volumes of alkali and acid used in the titration are now mixed together – this time without any indicator, which would contaminate the salt.

Making sodium chloride from sodium hydroxide solution and dilute hydrochloric acid

$$NaOH(aq) + HCl(aq) \rightarrow NaCl(aq) + H_2O(l)$$

The stages involved in this process are summarised below:

a) 25 ml of sodium hydroxide are added to a conical flask using a pipette. A few drops of indicator (phenolphthalein) are added.

b) The mixture is titrated with dilute hydrochloric acid from the burette. The acid is added until the indicator changes colour. (With phenolphthalein the change is pink to colourless.)

c) The titration is then repeated but without the indicator. The same volumes of each reactant are used to ensure that neither reactant is in excess.

d) The solution is then poured into an evaporating dish and evaporated to dryness. (It is not possible to crystallise sodium chloride by allowing a solution to cool because the solubility of sodium chloride is not very temperature dependant (see Figure 9.15.)

Figure 9.28
Sodium chloride is prepared by titrating sodium hydroxide with hydrochloric acid

Reacting a metal with an acid

This process is essentially the same as using an insoluble base. Excess metal is added to neutralise all the acid and the excess removed by filtration.

This method does not work for every metal. Metals lower than hydrogen in the **reactivity series** will not displace hydrogen from acids, so copper and silver salts cannot be made this way.

The reaction is usually:

metal + acid → metal salt + hydrogen

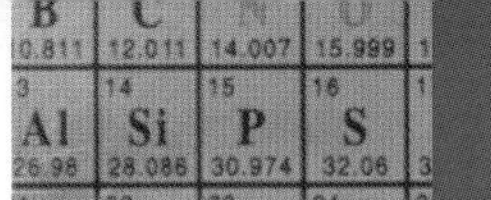

Figure 9.29
Iron(II) sulphate is prepared by adding iron filings to sulphuric acid

Did you know?

The acid/metal method does not work with nitric acid. Nitric acid is a powerful oxidising agent and will not liberate hydrogen. With nitric acid the reaction is:

metal + nitric acid → metal nitrate + nitrogen dioxide + water

Nitrogen dioxide is a toxic gas so this method is not normally used to produce metal nitrates.

Making iron(II) sulphate from iron and dilute sulphuric acid

$$H_2SO_4(aq) + Fe(s) \rightarrow FeSO_4(aq) + H_2(g)$$

The stages involved in this process are summarised below:

a) A beaker containing 50 ml of 2M sulphuric acid is heated and iron filings are slowly added. After each addition, the mixture is stirred until the iron has dissolved. Hydrogen gas is produced and the mixture bubbles quite a lot. (Other gases are also given off because there are impurities in the iron. These gases include hydrogen sulphide so the mixture has an unpleasant smell.) The solution turns pale blue-green. It takes about 6 g of iron to neutralise the acid completely.

b) Once that quantity has been added, all the sulphuric acid has been neutralised. The solution is filtered to remove the unreacted iron filings.

c) The solution is then boiled to remove some of the water. When the total volume is down to about 20 ml, the beaker is left to cool.

d) Green-blue crystals of iron(II) sulphate are produced as the solution evaporates.

NOTE
If weak acids are used in this method, the rate of reaction is very slow. This is because the key reaction is:

metal + hydrogen ions → metal ions + hydrogen

In this reaction the hydrogen ions are one of the reactants. In weak acids the hydrogen ion concentration will be very low so the rate of reaction will be slow. You can actually distinguish between strong and weak acids by the rate of their reaction with a suitable metal like magnesium.

Making an insoluble salt by precipitation

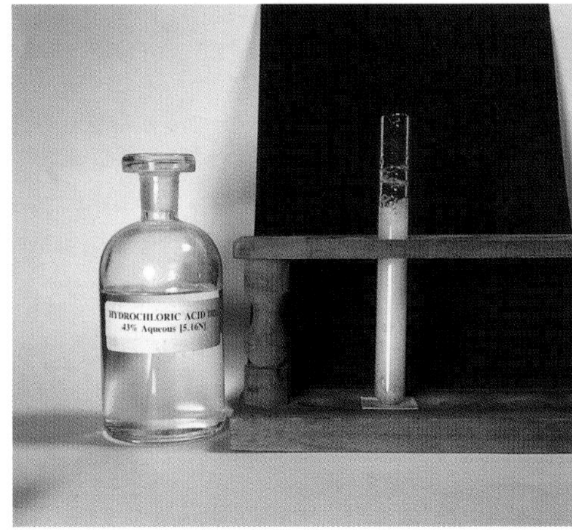

Figure 9.30
Marble chips dissolve more rapidly in dilute hydrochloric acid (a) than in dilute sulphuric acid (b)

As a general rule, insoluble salts cannot be prepared by reacting an insoluble base or a metal in dilute acid. A good example of this is shown in Figure 9.30. Marble chips (calcium carbonate) dissolve rapidly in hydrochloric acid but very slowly in sulphuric acid of the same concentration. This is because in sulphuric acid, the marble chip gets covered with a layer of insoluble calcium sulphate that reduces the effective surface area of the chip and slows down the reaction.

Insoluble salts like calcium sulphate are best prepared by **precipitation** reactions. In this process separate solutions each containing one of the ions present in the insoluble salt are mixed. The reaction is:

$$\text{metal ion}^{+}(aq) + \text{'acid' ion}^{-}(aq) \rightarrow \text{salt}(s)$$

Figure 9.31
Insoluble lead(II) iodide is made by mixing solutions of potassium iodide and lead nitrate

With this method there is no need to worry about using exactly the right amount of each reactant. The insoluble salt produced is filtered out. Any contaminating reactant can easily be removed by washing the insoluble salt with water. This process also removes the other soluble product formed in the reaction.

Making lead(II) iodide by precipitation

$$Pb(NO_3)_2(aq) + 2KI(aq) \rightarrow 2KNO_3(aq) + PbI_2(s)$$

The stages involved in the process are summarised below:

a) 10 ml of potassium iodide solution are mixed with an equal volume of lead(II) nitrate solution.

b) The resulting mixture is filtered. The potassium nitrate solution (and any excess reactant) is washed out with cold water.

c) Pure lead(II) iodide is left in the filter paper.

Making an anhydrous salt by direct combination of elements

This method can only be used to make salts where the **anion** (negative ion) is a single element. This means that it is particularly useful for making metal halides. Both aluminium and iron(III) chloride can be made this way. (It is not possible to make iron(III) chloride by dissolving iron in hydrochloric acid. The product of this reaction is iron(II) chloride.)

Because this method uses dry reactants, the salt produced is **anhydrous**.

Most salts contain water molecules locked inside the crystal structure. This water is called water of crystallisation. The amount of water of crystallisation is shown in the formula of the salt. For example copper(II) sulphate has 5 molecules of water to every 'molecule' of copper(II) sulphate and its formula is written: $CuSO_4.5H_2O$. In copper(II) sulphate, 36.0% of the crystal's mass is water. Other examples are:

$FeSO_4.7H_2O$ containing 45.3% water by mass.
$CaCl_2.6H_2O$ containing 49.3% water by mass.
$Na_2CO_3.10H_2O$ containing 62.9% water by mass.

Anhydrous salts contain no water of crystallisation. Some, like sodium chloride, only form anhydrous salts, others can exist in both forms. Hydrated copper(II) sulphate exists as blue crystals; anhydrous copper(II) sulphate is a white powder. If water is added to anhydrous copper(II) sulphate it reverts back to the blue, hydrated form. This is the familiar test for water.

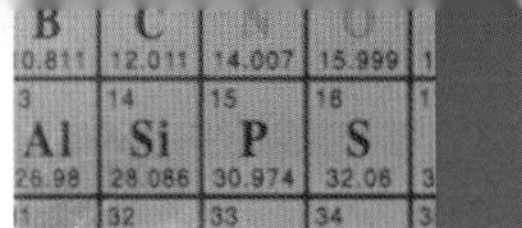

Making aluminium chloride by directly combining aluminium and chlorine

$$2Al(s) + 3Cl_2(g) \rightarrow 2AlCl_3(s)$$

The stages involved in this process are summarised below:

a) Aluminium turnings (or coarse powder) are placed in a tube. A suitable container is fixed to the end of the tube and chlorine gas is passed over the aluminium. (**CAUTION:** Chlorine is toxic and this method can *only* be done in a fume cupboard.)

b) The aluminium is heated in the stream of chlorine.

c) Anhydrous aluminium chloride is collected in the container.

Figure 9.32 *Aluminium chloride is prepared by passing chlorine gas over aluminium turnings*

Topic question

1 Outline a method for making reasonably pure samples of the following salts starting with the substance listed. You may also use sulphuric acid, hydrochloric acid or nitric acid. For each method give an equation.

All the salts are soluble in water except silver chloride.

a) Zinc sulphate starting with zinc metal.
b) Potassium chloride starting with a solution of potassium hydroxide.
c) Lead(II) nitrate starting with lead(II) carbonate. (Lead carbonate is not soluble in water.)
d) Magnesium sulphate starting with magnesium oxide. (Magnesium oxide is not soluble in water.)
e) Silver chloride starting with silver nitrate solution.

Summary

- Salts can be made by several processes:
  a) by reacting an acid with a base.
  b) by reacting an acid with a metal carbonate.
  c) by reacting an acid with a metal.
  d) by the direct combination of two elements to produce metal chlorides.

  Not every one of these processes can be used to produce every salt.
- Insoluble salts can be produced by precipitation.

9.7 Measuring the concentration of solutions

Co-ordinated	Modular
10.15	21 (14.5)

Doing a titration

The titration method that was used to prepare salts can be used to measure the concentration of a solution. The following method is used.

a) Using an accurate pipette, a volume of one of the reagents is transferred to a clean vessel (usually a conical flask).

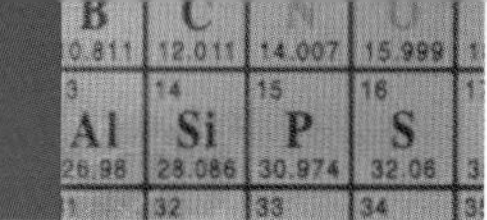

Figure 9.33
Carrying out a titration

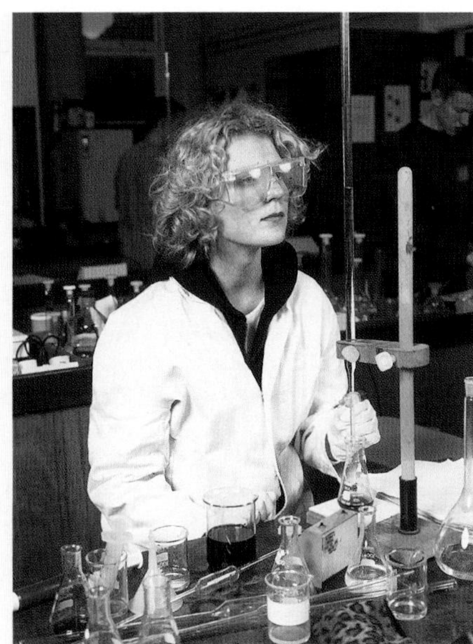

Did you know?

There are many indicators suitable for acid/base titrations. Litmus and Universal Indicator are not suitable as the colour change with these indicators is indistinct. The ones normally used are methyl orange (colour change: red (acid) to yellow (alkali)) and phenolphthalein (colour change: colourless (acid) to pink (alkali)).

b) A few drops of indicator solution are added to the flask. (It is easier to see the endpoint if the colour is not too intense. If the colour seems rather pale it is always possible to add more later.)
c) The other reagent is placed in a burette. The liquid must be run through the burette until the tap and the jet are filled. (The burette does not have to be filled to the 0.0 mark.) The exact initial reading is taken. Figure 9.34 shows how to read a burette.

Always:

i) Get the eye level with the meniscus.
ii) Read to the bottom of the meniscus.
ii) Record the reading.

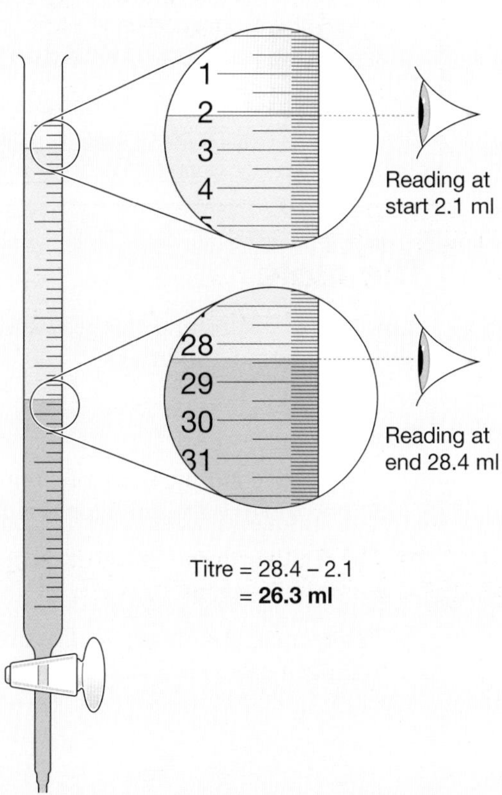

Figure 9.34
The correct way to read a burette

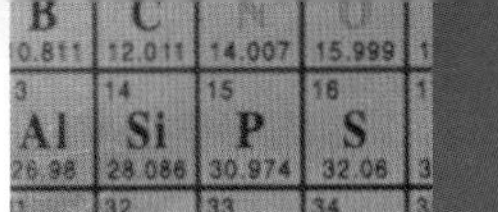

d) The liquid in the burette is added carefully to the contents of the titrating flask. The flask must be swirled to make sure the contents are well mixed.

e) When the endpoint is reached, the burette is read again.

f) The titre is calculated by subtracting the first reading from the second reading.

g) The first titre value will not be very accurate – it is a 'ranging shot' – so the entire process is repeated. This time when the endpoint is near, the liquid from the burette is added drop by drop. (It should be possible to get the titration accurate to the nearest drop (about 0.05 ml).)

h) The titration flask is washed out with distilled water and the titration is repeated. Repeats are made until two or more values for the titre are obtained that are the same to within 0.05 ml.

i) Only these accurate values should be used in the calculation.

Titrations can be used in this way to find the concentrations of one of the reactants if the concentration of the other is known.

Did you know?

The titration method can be used for determining concentration in any reaction where there is a detectable change in the mixture as the reaction is completed. In acid/base titrations the change is detected by the addition of a suitable indicator. Titrations can be used in:

- redox reactions involving transition metals where one of the reagents changes colour.
- reactions where iodine is either liberated or used (the colour change can be enhanced by the use of starch as an indicator).
- reactions between the silver ion $Ag^+(aq)$ and halide ions. Silver halide is precipitated. The end point is determined using a few drops of potassium chromate which goes brick red when all the halide ions have been used and free silver ions are present.

Other methods for detecting the end point include monitoring the electrical conductivity of the solution, measuring the pH of the solution using a meter rather than indicators and measuring the temperature change during a reaction.

The mole

A mole of any substance is the **relative atomic mass** (elements) or the **relative formula mass** (compounds) in grams.

Examples

a) The relative atomic mass of sodium is 23. This can be written as A_r (Na) = 23. So one mole of sodium atoms is 23 grams.

b) The relative atomic mass of oxygen is 16. So A_r (O) = 16. One mole of oxygen atoms is 16 grams.

c) The relative atomic mass of hydrogen is 1. So A_r (H) = 1. Three moles of hydrogen atoms is 3 × 1 grams = 3 grams.

d) For sulphur, A_r (S) = 32. So 0.3 moles of sulphur atoms is 0.3×32 grams = 9.6 grams.

e) Oxygen gas exists as O_2 molecules. The relative formula mass of oxygen gas (written as $M_r(O_2)$) is $2 \times 16 = 32$. One mole of oxygen gas (or oxygen molecules) is 32 grams.

NOTE: One mole of oxygen molecules is *not* the same as one mole of oxygen atoms. It is necessary to state *exactly* what the substance is. To say '1 mole of oxygen' is unclear – does it mean atoms or molecules?

f) One mole of hydrogen gas (or hydrogen molecules H_2) is $2 \times 1 = 2$. So $M_r(H_2) = 2$ and one mole of hydrogen gas = 2 grams.

g) The formula of sodium hydroxide is NaOH. One molecule of sodium hydroxide contains:

- 1 sodium atom A_r (Na) = 23
- 1 oxygen atom A_r (O) = 16
- 1 hydrogen atom A_r (H) = 1

So $M_r(\text{NaOH}) = (1 \times 23) + (1 \times 16) + (1 \times 1) = 40$.

One mole of sodium hydroxide has a mass of 40 grams
and 0.6 moles of sodium hydroxide has a mass of $0.6 \times 40 = 24$ grams.

h) The formula of sulphuric acid is H_2SO_4. One molecule of sulphuric acid contains:

- 2 hydrogen atoms A_r (H) = 1
- 1 sulphur atom A_r (S) = 32
- 4 oxygen atoms A_r (O) = 16

So $M_r(H_2SO_4)$ is $(2 \times 1) + (1 \times 32) + (4 \times 16) = 98$

One mole of sulphuric acid has a mass of 98 grams and 1.5 moles of sulphuric acid has a mass of $1.5 \times 98 = 147$ grams

This relationship can also be worked in the opposite direction.

$M_r(Na_2S)$ is $(2 \times 23) + (1 \times 32) = 78$.

78 grams of sodium sulphide is 1 mole
so 1 gram of sodium sulphide is $1/78$ moles
and 234 grams of sodium sulphide is $234/78 = 3$ moles

The general rule

$$\text{Number of moles} = \frac{\text{Mass/g}}{A_r \text{ or } M_r}$$

Using the mole in concentrations

The concentration of aqueous solutions is normally expressed in moles per cubic decimetre (mol dm^{-3}). (Notice that the symbol for 'mole' is 'mol'.) This is also called the molarity of the solution and has the symbol M.

So a solution containing 3 moles of solute in 1 cubic decimetre of solution has a concentration of 3 mol dm^{-3}. It is a 3 molar solution (3M solution).

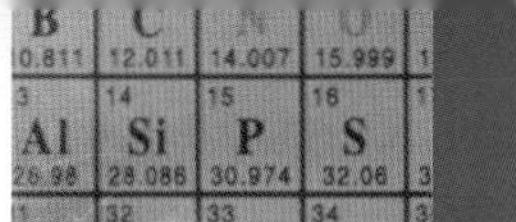

Examples

a) 0.2 mol of H_2SO_4 is present in 1 dm^3 of solution.
The concentration is 0.2 mol dm^{-3} or 0.2M.

b) 500 ml of a solution contains 0.8 mol of NaOH.
So 1000 ml (1 dm^3) of solution would contain:

$$\frac{1000}{500} \times 0.8 = 1.6 \text{ mol of NaOH}$$

The concentration is 1.6 mol dm^{-3} (1.6M).

c) A solution of potassium chloride (KCl) has a concentration of 0.3M. What mass of KCl is present in 10 ml of this solution? [A_r(K) = 39, A_r(Cl) = 35.5]

M_r(KCl) = (1 × 39) + (1 × 35.5) = 74.5

So 1 mole is 74.5 grams.

A 0.3M solution contains 0.3 mole in 1 dm^3 of solution

0.3 mole = 0.3 × 74.5 g
= 22.35 g

So 1 dm^3 (1000 ml) of solution contains 22.35 g
and 10 ml contains:

$$22.35 \times \frac{10}{1000}$$

$$= 0.2235 \text{ g}$$

The general rule

v ml of a ***M*** molar solution (***M*** mol dm^{-3}) contains

$$\frac{v \times M}{1000} \times M_r \text{ g of solute}$$

Calculating concentrations from titrations

If the concentration of the solution of one of the reactants is known, a titration can be used to determine the concentration of the other reactant.

Examples

a) In a titration, 25 ml of NaOH were neutralised by 26.3 ml of 0.1M HCl. What is the concentration of the NaOH?

i) The equation is:

$$NaOH + HCl \rightarrow NaCl + H_2O$$

ii) The base:

Call the concentration of the NaOH **c** mol dm^{-3}

25 ml of the solution contain:

$$\frac{25 \times c}{1000} \text{ moles of NaOH}$$

iii) The acid:

26.3 ml of 0.1M HCl contain:

$$\frac{26.3 \times 0.1}{1000} \text{ moles of HCl}$$

iv) Equating the concentrations

From the equation, 1 mole of NaOH is neutralised by 1 mole of HCl.

So:

$$\frac{25 \times \mathbf{c}}{1000} = \frac{26.3 \times 0.1}{1000}$$

Cancelling the 1000s gives:

$$25 \times \mathbf{c} = 26.3 \times 0.1$$

$$\mathbf{c} = \frac{26.3 \times 0.1}{25} = 0.1052 \text{ mol dm}^{-3}$$

The titration is only accurate to ± 0.05 ml which is a percentage error of ± $(^{0.05}/_{26.3}) \times 100 = 0.2\%$ so it is only possible to quote the concentration to this level of accuracy.

The concentration is **0.105 mol dm^{-3}**.

b) In a titration, 20 ml of 0.24M KOH were neutralised by 11.5 ml of H_2SO_4. What is the concentration of the H_2SO_4 ?

i) The equation is:

$$2KOH + H_2SO_4 \rightarrow K_2SO_4 + 2H_2O$$

ii) The base:

20 ml of the solution contain:

$$\frac{20 \times 0.24}{1000} \text{ moles of KOH}$$

iii) The acid:

Call the concentration of the H_2SO_4 ***a*** mol dm^{-3}

11.5 ml of ***a*** M H_2SO_4 contain:

$$\frac{11.5 \times \boldsymbol{a}}{1000} \text{ moles of } H_2SO_4$$

iv) Equating the concentrations:

From the equation, 2 moles of KOH are neutralised by 1 mole of H_2SO_4.

So 1 mole of KOH is neutralised by 0.5 mole of H_2SO_4

so:

$$\frac{20 \times 0.24}{1000} \text{ are neutralised by: } 0.5 \times \frac{20 \times 0.24}{1000} \text{ moles of } H_2SO_4$$

Equating the number of moles of H_2SO_4

$$= 0.5 \times \frac{20 \times 0.24}{1000} = \frac{11.5 \times \boldsymbol{a}}{1000}$$

Cancelling the 1000s gives:

$$0.5 \times 20 \times 0.24 = 11.5 \times \boldsymbol{a}$$

$$\boldsymbol{a} = \frac{0.5 \times 20 \times 0.24}{11.5} = 0.209 \text{ mol dm}^{-3}$$

The concentration is **0.209 mol dm^{-3}**.

The general rule

c) In a titration, $\boldsymbol{V_A}$ ml of $\boldsymbol{M_A}$ M acid ($\boldsymbol{M_A}$ mol dm^{-3}) were neutralised by $\boldsymbol{V_B}$ ml of $\boldsymbol{M_B}$ M base ($\boldsymbol{M_B}$ mol dm^{-3}).

i) The balanced chemical equation is:

a ACID + **b** BASE → products

a moles of acid are neutralised by **b** moles of base

ii) The base:

$\boldsymbol{V_B}$ ml of the solution contain:

$$\frac{\boldsymbol{V_B} \times \boldsymbol{M_B}}{1000} \text{ moles of base}$$

iii) The acid:

$\boldsymbol{V_A}$ ml of the solution contain:

$$\frac{\boldsymbol{V_A} \times \boldsymbol{M_A}}{1000} \text{ moles of acid}$$

iv) Equating the concentrations:

From the equation, **b** mole of base are neutralised by **a** mole of acid.

So 1 mole of base is neutralised by $^{\mathbf{a}}/_{\mathbf{b}}$ mole of acid.

So:

$$\frac{\boldsymbol{V_B} \times \boldsymbol{M_B}}{1000} \text{ of base are neutralised by: } \frac{\mathbf{a}}{\mathbf{b}} \times \frac{\boldsymbol{V_B} \times \boldsymbol{M_B}}{1000} \text{ moles of acid}$$

Equating the number of moles of acid:

$$\frac{\mathbf{a}}{\mathbf{b}} \times \frac{\boldsymbol{V_B} \times \boldsymbol{M_B}}{1000} = \frac{\boldsymbol{V_A} \times \boldsymbol{M_A}}{1000}$$

Cancelling the 1000s gives:

$$\frac{\mathbf{a}}{\mathbf{b}} \times \boldsymbol{V_B} \times \boldsymbol{M_B} = \boldsymbol{V_A} \times \boldsymbol{M_A}$$

which can be rearranged to:

$$\frac{\boldsymbol{V_B} \times \boldsymbol{M_B}}{\mathbf{b}} = \frac{\boldsymbol{V_A} \times \boldsymbol{M_A}}{\mathbf{a}}$$

Topic questions

1 In a titration the following results are obtained:

burette reading	1st titration	2nd titration	3rd titration	4th titration
2nd reading	21.85	21.95	22	21.85
1st reading	0.15	0.4	0.55	0.3
titre				

Calculate the titre values and decide the correct value to use in a calculation.

In the following calculations use these values for the relative atomic masses.
H = 1; C = 12; N = 14; O = 16; Na = 23; S = 32; Cl = 35.5; K = 39; Ca = 40

2 What is the mass of:
a) 1 mol of sulphur atoms?
b) 2 mol of chlorine molecules?
c) 0.5 mol of water molecules?
d) 1.35 mol of hydrochloric acid molecules?

3 How many moles are in:
a) 46 g of sodium atoms?
b) 3.9 g of potassium atoms?
c) 1.98 g of water molecules?
d) 0.803 g of hydrogen chloride molecules?

4 What is the molar concentration of the following solutions?
a) 2 g of sodium hydroxide in 1 litre of solution.
b) 3.15 g of nitric acid in 500 ml of solution.
c) 0.49 g of sulphuric acid in 2 litres of solution.

5 How many grams of solute are present in the following solutions?
a) 100 ml of 2M sodium chloride solution.
b) 0.25 ml of 1M potassium carbonate solution.
c) 1.5 litres of 0.01M sulphuric acid.
d) 52.5 ml of 0.15M ammonia solution. (Ammonia is NH_3).

6 Calculate the missing values in the following:
a) 20 ml of 0.1M HCl(aq) neutralise 22.1 ml of _____ M NaOH(aq).
b) 25 ml of 0.05M H_2SO_4(aq) neutralise ______ ml of 0.15M KOH(aq).
c) ______ ml of 0.12M HNO_3(aq) neutralise 15 ml of 0.1M $Ca(OH)_2$(aq)

Summary

- A **mole** of any substance is the relative atomic mass (or relative formula mass) of that substance in grams.
- The number of moles of a substance
$$= \frac{\text{mass of substance (in grams)}}{\text{relative atomic (or formula) mass}}$$
- The **molarity** of a solution is the number of moles of solute in 1 dm^3 (litre) of solution.
- The number of moles of solute in **v** ml of a **M** molar solution $= \frac{\mathbf{v} \times \mathbf{M}}{1000}$
- The number of grams of solute in **v** ml of a **M** molar solution
$$= \frac{\mathbf{v} \times \mathbf{M}}{1000} \times M_r\text{g of solute}$$
- In a neutralisation reaction if **a** moles of acid **A** exactly neutralise **b** moles of base **B** and $\mathbf{V_A}$ ml of a $\mathbf{M_A}$ molar solution of acid **A** exactly neutralise $\mathbf{V_B}$ ml of a $\mathbf{M_B}$ molar solution of base **B** then:
$$\frac{(\mathbf{V_A} \times \mathbf{M_A})}{\mathbf{a}} = \frac{(\mathbf{V_B} \times \mathbf{M_B})}{\mathbf{b}}$$

Examination questions

1 This question is about the water cycle.
The water cycle can be described in nine sentences.
In the cycle below, five of the sentences are missing.
These five missing sentences are given in the list below in the **wrong** order.

A These rise further and cool.
B This heat evaporates the water producing water vapour.
C As this happens, clouds are formed.
D This water vapour rises into the atmosphere.
E This cooling causes water droplets to form rain.

Complete the description of the water cycle by writing the letters **A, B, C, D** and **E** in the boxes in the correct order.

'The Water Cycle'

- The water cycle starts with water in rivers, lakes and oceans.
- This water is warmed by the heat of the sun.
- ☐
- ☐
- ☐
- ☐
- ☐
- This falls from the sky onto seas and land.
- The water cycle begins again. *(3 marks)*

2 The label shows the ions present in the bottle of spring water. This water is *temporarily* hard.

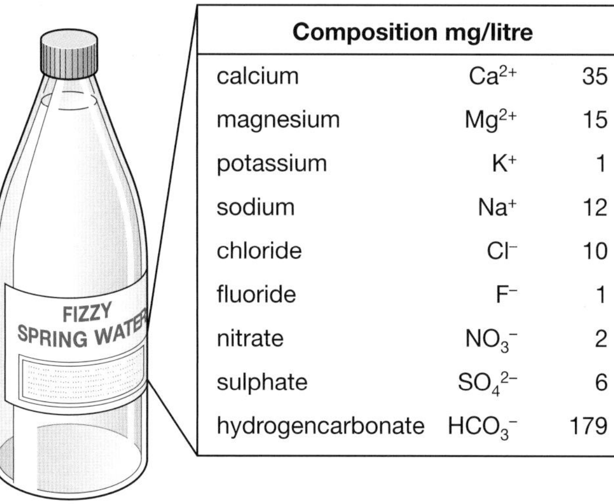

Composition mg/litre		
calcium	Ca^{2+}	35
magnesium	Mg^{2+}	15
potassium	K^+	1
sodium	Na^+	12
chloride	Cl^-	10
fluoride	F^-	1
nitrate	NO_3^-	2
sulphate	SO_4^{2-}	6
hydrogencarbonate	HCO_3^-	179

a) Name the compound that would be present in the greatest amount if this water were evaporated to dryness. *(2 marks)*

b) i) What is hard water? *(2 marks)*
ii) State one advantage of hard water. *(1 mark)*

c) Describe an experiment that would show that this water is *temporarily* hard. *(4 marks)*

d) This hard water may be softened as shown.

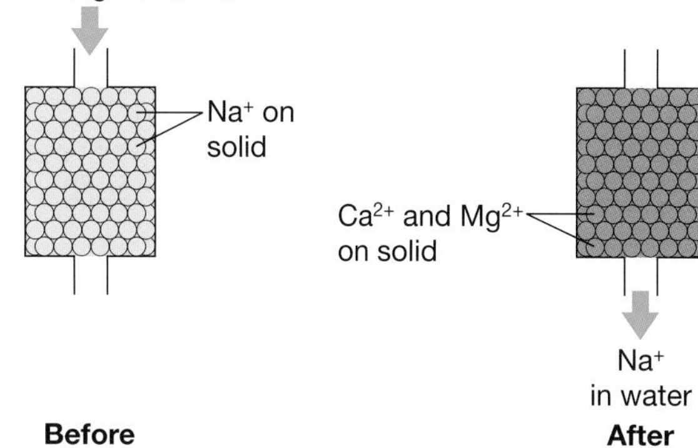

What name is given to this process? *(1 mark)*

3 a) Explain what is meant by
i) a **saturated** solution *(2 marks)*
ii) a **saturated** hydrocarbon *(1 mark)*

b) The table shows some data for the solubility of ammonium chloride in water at different temperatures.

Temperature (°C)	Solubility (g per 100 g water)
0	29.4
20	37.2
40	40.5
60	55.2
80	65.6
100	77.3

Select a suitable scale for each axis and plot the data on a grid like the one shown.
Draw a smooth curve to show how the solubility of ammonium chloride in water changes with temperature, allowing for any anomalous point.

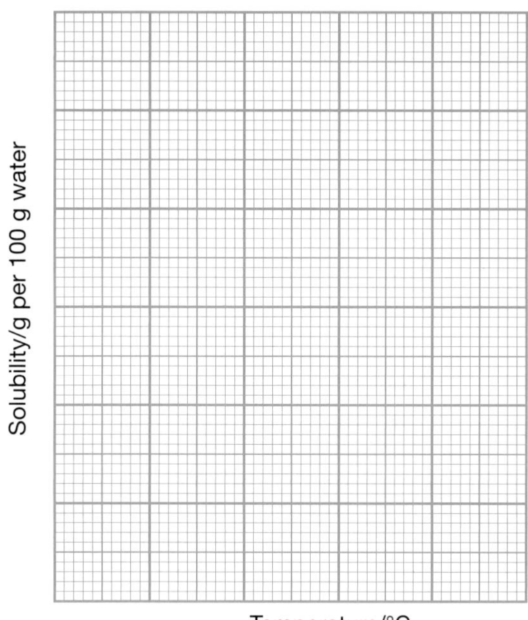

(3 marks)

c) Use your graph to answer the following questions.

i) What is the solubility of ammonium chloride at 90°C? *(1 mark)*

ii) What is the lowest temperature at which 50 g of ammonium chloride dissolves completely in 100 g of water? *(1 mark)*

iii) What mass of ammonium chloride crystals would be obtained if a saturated solution of ammonium chloride, prepared using 50 g of water, was cooled from 100°C to 30°C? *(2 marks)*

4 Ethanoic acid, CH_3COOH, forms a *weak acid* when added to water. Some reactions of ethanoic acid are shown.

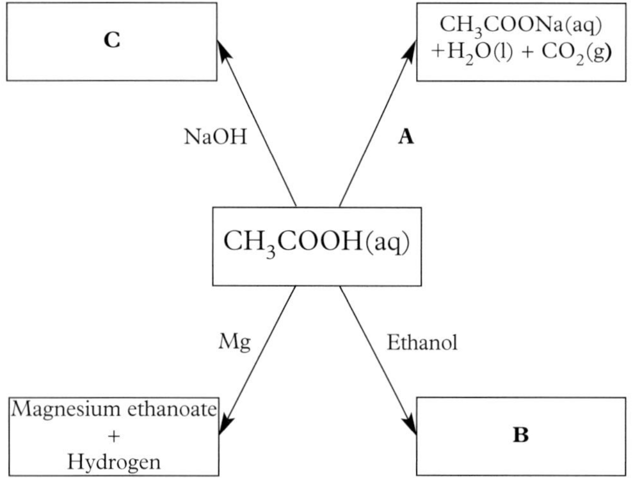

a) Explain what is meant by a *weak acid*. *(2 marks)*

b) Name the substance **A** that is added to ethanoic acid. *(1 mark)*

c) Substance **B** is formed when ethanoic acid reacts with ethanol. What type of substance is **B**? *(1 mark)*

d) Draw a displayed structural formula for salt **C**. *(1 mark)*

e) Write a balanced chemical equation for the reaction between magnesium and ethanoic acid. *(2 marks)*

5 Salts can be made by neutralising acids. A student followed these instructions to make common salt, sodium chloride.

a)

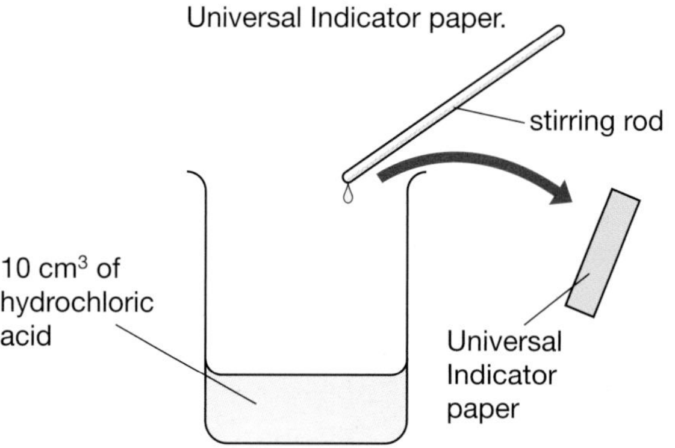

What colour would the Universal Indicator paper go? *(1 mark)*

b)

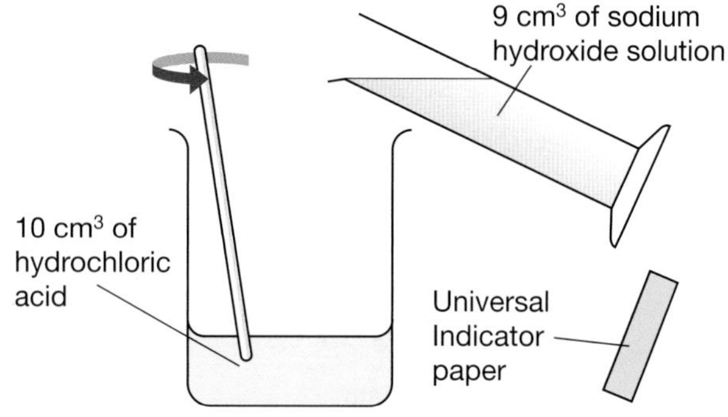

The colour of the Universal Indicator paper was the same as in (a). Why? *(1 mark)*

c)

Instruction 3 – Now add more sodium hydroxide solution a drop at a time. Stir the solution each time and test with Universal Indicator paper.

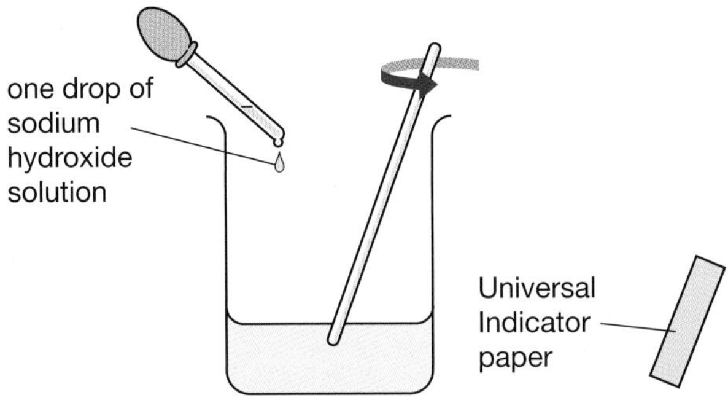

i) The Universal Indicator paper showed that the solution was now alkaline. What should be done to make the solution neutral? *(1 mark)*

ii) When the solution was neutral, what colour and pH was shown by the Universal Indicator paper? *(2 marks)*

d) The equation for the reaction is:

$$NaOH(aq) + HCl(aq) \rightarrow NaCl(aq) + H_2O(l)$$

i) What is the chemical formula of hydrochloric acid? *(1 mark)*

ii) Sodium chloride crystals often form in the neutral solution. Explain how. *(2 marks)*

iii) Sodium chloride for use in the home is not prepared by this reaction. Explain why. *(2 marks)*

6 An oven cleaner solution contained sodium hydroxide. A 25.0 cm^3 sample of the oven cleaner solution was placed in a flask. The sample was titrated with hydrochloric acid containing 73 g/dm^3 of hydrogen chloride, HCl.

a) Describe how this titration is carried out. *(3 marks)*

b) Calculate the concentration of the hydrochloric acid in mol/dm^3.
Relative atomic masses: H 1; Cl 35.5 *(2 marks)*

c) 10.0 cm^3 of hydrochloric acid were required to neutralise the 25.0 cm^3 of oven cleaner solution.

i) Calculate the number of moles of hydrochloric acid reacting. *(2 marks)*

ii) Calculate the concentration of sodium hydroxide in the oven cleaner solution in mol/dm^3. *(2 marks)*

7 A student carried out a titration to find the concentration of a solution of sulphuric acid. 25.0 cm^3 of the sulphuric acid solution was neutralised exactly by 34.0 cm^3 of a potassium hydroxide solution of concentration 2.0 mol/dm^3. The equation for the reaction is

$$2KOH(aq) + H_2SO_4(aq) \rightarrow K_2SO_4(aq) + 2H_2O(l)$$

a) Describe the experimental procedure for the titration carried out by the student. *(4 marks)*

b) Calculate the number of moles of potassium hydroxide used. *(2 marks)*

c) Calculate the concentration of the sulphuric acid in mol/dm^3. *(3 marks)*

Chapter 10

Detection and identification

Key terms

flame test • gas chromatography • **ion** • mass spectrometry • **molar** • nuclear magnetic resonance • **precipitate** • spectroscopy • **thermal decomposition**

10.1	
Co-ordinated	Modular
10.16	22 (15.4)

Laboratory methods

It is possible to identify almost any substance by chemical methods. The tests mentioned in this section are a sample of the methods that can be used.

Identifying positive ions

Flame tests

In a **flame test** the substance to be tested is placed in a watch glass. A clean platinum wire is placed in the substance and the wire is then touched to the edge of a blue Bunsen flame. Some metal **ions** will colour the flame. Figure 10.1 shows the colours obtained by different elements. These are summarised in Figure 10.2.

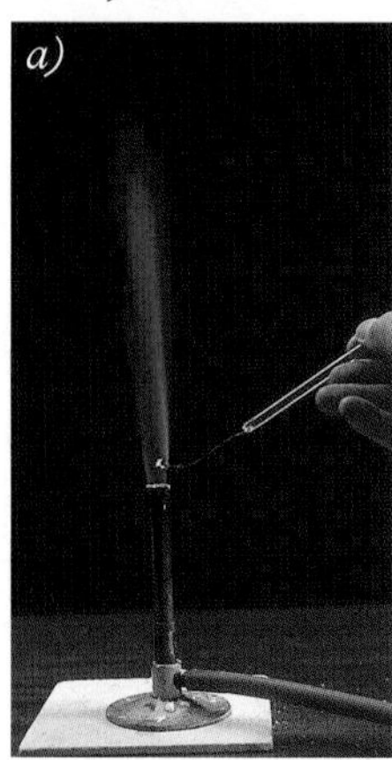

a)

b)

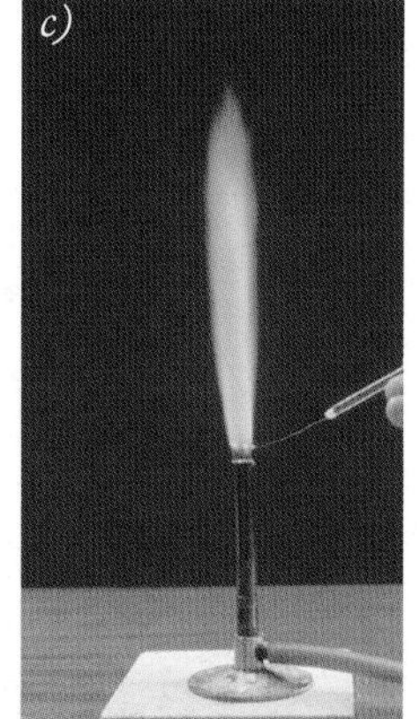

c)

d)

e)

Figure 10.1 *Certain elements produce certain coloured flames. The elements shown here are a) lithium, b) sodium, c) calcium, d) potassium and e) barium*

Metal ion	Flame colour
lithium (Li^{+})	bright red
sodium (Na^{+})	golden yellow
potassium (K^{+})	lilac
calcium (Ca^{2+})	brick red
barium (Ba^{2+})	apple green

Figure 10.2 *The flame colours produced by the five elements from Figure 4.1*

Other metals like copper and lead also colour the flame. Do not confuse the flame colour in this test with the colour of the flame when the element is burned in oxygen. In some cases the colour is the same (e.g. sodium and calcium), in others it is not. Magnesium metal burns with an intense white flame in oxygen but magnesium compounds do not colour a flame.

Did you know?

When a metal ion is heated, some of the electrons gain enough energy to move into a different electron level. On cooling slightly, the electron drops back to its original level. The energy given out in the drop is often given out as light energy. For some metals the energy given out is in the infrared or ultraviolet parts of the electromagnetic spectrum so the flame has no visible colour.

Adding sodium hydroxide solution

Two or 3 ml of a solution of the substance being tested are placed in a test tube. **Molar** sodium hydroxide is added slowly until it is in excess. Figure 10.3 shows the effect of sodium hydroxide solution on different ions.

Figure 10.3
The effect of sodium hydroxide solution on different ions

Ion present	Effect of adding sodium hydroxide solution	
	A few drops of the solution	An excess of the solution
aluminium (Al^{3+})	white, gelatinous precipitate	precipitate re-dissolves
calcium (Ca^{2+})	white precipitate	no change to precipitate
magnesium (Mg^{2+})	white precipitate	no change to precipitate
copper(II) (Cu^{2+})	blue-green, gelatinous precipitate	no change to precipitate
iron(II) (Fe^{2+})	green-grey, gelatinous precipitate	no change to precipitate
iron(III) (Fe^{3+})	red-brown gelatinous precipitate	no change to precipitate
ammonium (NH_4^+)	ammonia gas given off when heated	

Test for ammonia
Ammonia has a distinctive, pungent odour. The gas turns damp red litmus paper blue

Figure 10.4
The precipitates of the following hydroxides: a) aluminium hydroxide, b) magnesium hydroxide, c) copper(II) hydroxide, d) iron(III) hydroxide

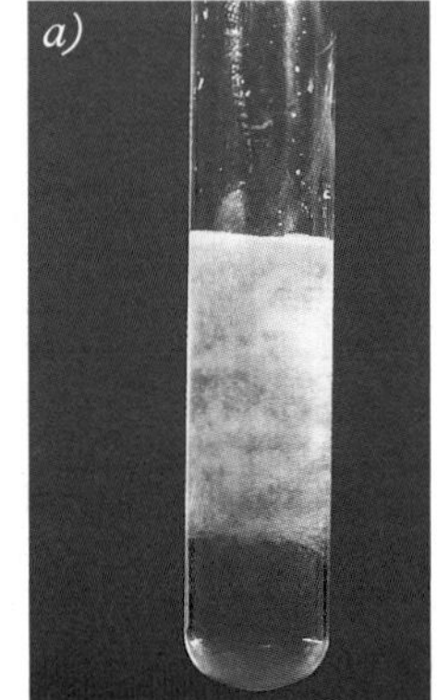

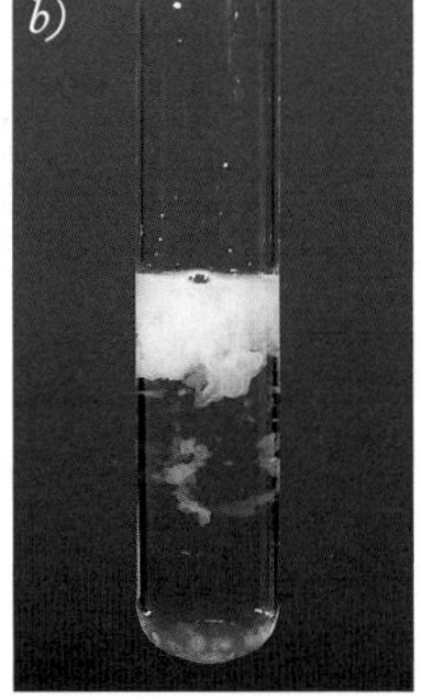

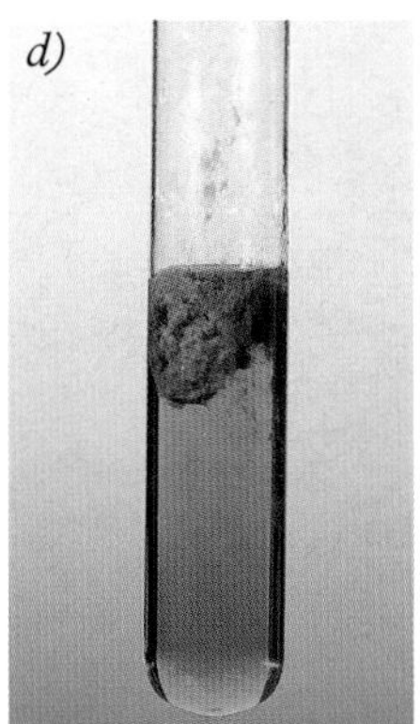

In each of the reactions with metal ions, the **precipitate** is the metal hydroxide. For the 2+ ions, the reaction is:

$$M^{2+}(aq) + 2NaOH(aq) \rightarrow 2Na^+(aq) + M(OH)_2(s)$$

For the 3+ ions, the reaction is:

$$M^{3+}(aq) + 3NaOH(aq) \rightarrow 3Na^+(aq) + M(OH)_3(s)$$

Aluminium hydroxide re-dissolves in sodium hydroxide to form sodium aluminate solution.

With the ammonium ion, the reaction is:

$$NH_4^+(aq) + NaOH(aq) \rightarrow NH_3(g) + H_2O(l) + Na^+(aq)$$

Identifying negative ions

The standard tests for the common negative ions are shown in Figure 10.5.

Figure 10.5
The standard tests for the common negative ions

Negative ion present	Test	Positive result
carbonate (CO_3^{2-})	add dilute acid	carbon dioxide gas given off
chloride (Cl^-)	add dilute nitric acid *then* add silver nitrate solution	white precipitate
bromide (Br^-)	add dilute nitric acid *then*	cream coloured precipitate
iodide (I^-)	add silver nitrate solution	yellow coloured precipitate
sulphate (SO_4^{2-})	add dilute hydrochloric acid *then* add barium chloride solution	white precipitate
nitrate (NO_3^-) (see note below)	add sodium hydroxide solution *then* add aluminium powder *and* heat the mixture	ammonia gas given off

Note: The nitrate test can't be used if the ammonium ion is present. It is necessary to remove the ammonium ion before testing for the nitrate ion. This can be done by heating the substance with sodium hydroxide until all the ammonia has been evolved *then* adding aluminium powder and heating.

The reactions involved are:

carbonate $$CO_3^{2-}(aq) + 2H^+(aq) \rightarrow H_2O(l) + CO_2(g)$$

chloride $$Cl^-(aq) + Ag^+(aq) \rightarrow AgCl(s)$$

bromide $$Br^-(aq) + Ag^+(aq) \rightarrow AgBr(s)$$

iodide $$I^-(aq) + Ag^+(aq) \rightarrow AgI(s)$$

sulphate $$SO_4^{2-}(aq) + Ba^{2+}(aq) \rightarrow BaSO_4(s)$$

Heating substances

Many substances change when heated. These changes can be used to help in identification. Copper(II) carbonate (which is a green powder) is **thermally decomposed** when heated to produce a black powder and carbon dioxide gas.

$$CuCO_3(s) \rightarrow CuO(s) + CO_2(g)$$

Figure 10.6
When copper(II) carbonate is heated, copper oxide and carbon dioxide are formed

Zinc carbonate (a white powder) also decomposes when heated to produce white zinc oxide and carbon dioxide gas. The zinc oxide produced is yellow when hot but turns white again when it cools.

$$ZnCO_3(s) \rightarrow ZnO(s) + CO_2(g)$$

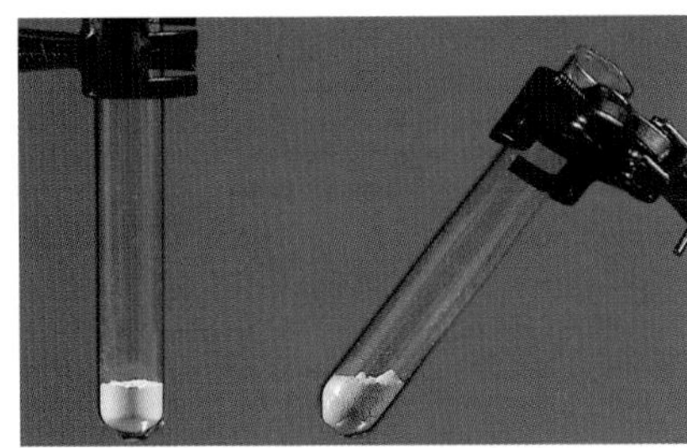

Figure 10.7
Zinc oxide is yellow when hot and white when cold

Summary

- **Flame tests** can be used to identify many positive ions.
- Many positive ions can also be identified by their reaction with sodium hydroxide solution.
- Various tests can be used to identify negative ions.

10.2 Instrumental methods

Co-ordinated	Modular
10.16	22 (15.5)

Laboratory methods of identification are slow and need relatively large amounts of material. In contrast, instrumental methods are very quick, accurate and sensitive, and often need only very small amounts of material. The instruments used are, however, very expensive and are only suitable when a lot of testing has to be done.

Using modern electronics and computer technology, instrumental methods have become very efficient. This makes them ideal for industrial use. Steel samples from the furnace can be analysed in minutes. The method used is **spectroscopy**.

Using this method, the accurate analysis is known before the steel is poured from the furnace (see Chapter 8). In the past, using laboratory analysis, the steel had to be solidified then samples taken and analysed. The process could take several days and required very skilled analytical chemists to carry out the work. If the composition was wrong, the steel would have to be re-melted and its composition changed. Re-melting the steel used a lot of energy and was very expensive.

Modern instrumental methods can also keep a rapid and accurate check on the environment. Oil tankers sometimes discharge oil residues from their tanks into the sea causing pollution. Using a combination of **gas chromatography** and **mass spectrometry** it is possible to analyse the contents of oil slicks with such accuracy that the tanker that caused the slick can be positively identified.

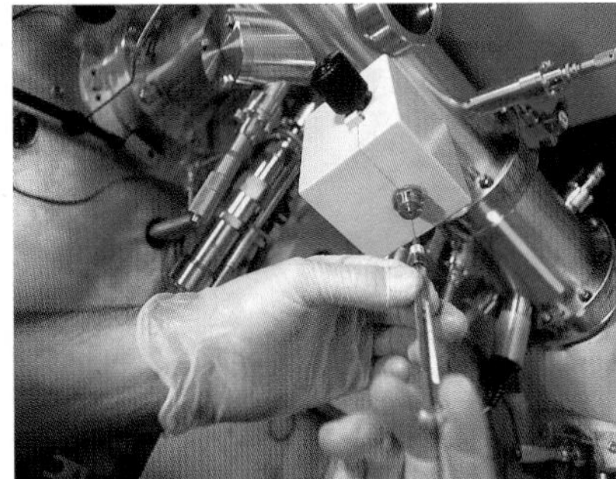

Figure 10.8
A gas chromatograph in use (left)

Figure 10.9
Using a mass spectrometer

All instrumental methods work by measuring a specific property of an element or compound. Figure 10.10 lists a number of instrumental methods and the scientific principle behind them.

Figure 10.10 *Some modern instrumental methods of analysis*

Method	Used to identify:	Scientific principle used
nuclear magnetic resonance (NMR)	organic compounds	The hydrogen atoms in organic compounds behave like small magnets. In a magnetic field they absorb electromagnetic energy. The frequency of the energy absorbed depends on what the hydrogen atom is connected to. The presence of functional groups like $-CH_3$, $=CH_2$, –OH etc. can be identified.
mass spectrometry	elements (and their isotopes)	In an electric field, positive ions will move towards the negative electrode. The amount of movement depends on the mass of the ion. (Lighter ions move more.) Measuring the deflection of an ion allows its mass to be determined.
emission spectroscopy	elements	When elements are heated, electrons can get enough energy to move to a higher energy level. When the electrons go back to the lower level they give out this energy as light (or infrared or ultraviolet energy) of a particular frequency. Different elements give out different frequencies. By identifying the frequencies, the elements present can be determined. This is what colours the flames in a flame test.
absorption spectroscopy	elements	This uses the same principle as emission spectroscopy. In this case energy is absorbed to move an electron to a higher level. The spectrum has dark bands at the frequencies absorbed.
infrared spectroscopy	compounds (mainly organic)	This method uses absorption spectroscopy in the infrared region. In this method, energy is absorbed by bonds in the molecules and not electrons in the atoms. Different types of bond absorb energy at different frequencies. So C–H, C = C, C–O and C = O bonds can be identified.
gas chromatography	volatile substances	The method is similar to paper chromatography. In gas chromatography the substance is passed along a tube (instead of a piece of paper) in a carrier gas (instead of a solvent like water or ethanol). The length of time it takes to get to the end of the tube is used to identify the substance.

These methods have been developed over many years. Spectroscopy was first developed by two German scientists (one was Robert Wilhelm von Bunsen – best known for his invention of the laboratory burner) in the 1850s. Mass spectrometry was developed in the 1920s and **nuclear magnetic resonance** (NMR) in the 1950s.

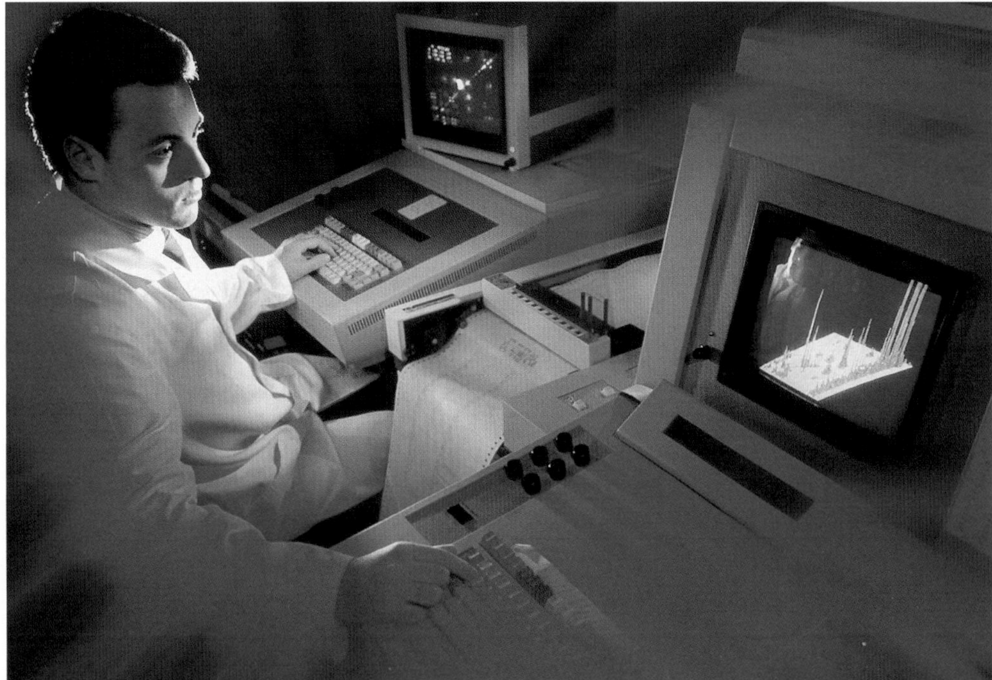

Figure 10.11 *A modern NMR machine*

The development of these methods to the sophisticated instruments used today is a result of the fact that the methods have been so useful. They have increased the speed and accuracy of analytical methods. They have resulted in many benefits to society: increased efficiency in industry, more accurate forensic analysis in the fight against crime, better diagnostic methods in medicine etc.

Topic questions

1 In a flame test, what element would be present if the flame had the following colours?
   a) Apple green.
   b) Golden yellow.
   c) Brick red.

2 Sodium hydroxide solution is added to a solution of a metal ion. Which metal ion is present if:
   a) a red-brown precipitate forms?
   b) a blue-green precipitate forms?
   c) a white precipitate forms which re-dissolves when excess sodium hydroxide is added?

3 Give full details of the test you would use for the ammonium ion.

4 Give full details of the test you would use for the carbonate ion.

5 Describe the test for the nitrate ion. What precautions must you take to ensure you don't get a false result?

6 Describe the test for the halide ions. How could you tell which halide was present?

7 Which instrumental method(s) would you use to do the following?
   a) Detect the carboxylic acid functional group (COOH) in a compound.
   b) Analyse a sample of petrol to find out which alkanes were in it.
   c) Prove that chlorine gas contains 75% ^{35}Cl and 25% ^{37}Cl.

Summary

- **Instrumental methods**, including spectroscopy, infrared spectroscopy, gas chromatography, mass spectrometry and nuclear magnetic resonance, can be used to identify elements and compounds.

Examination questions

1 A student performed some chemical reactions on **two** compounds.

a) Reactions starting with compound **S**.

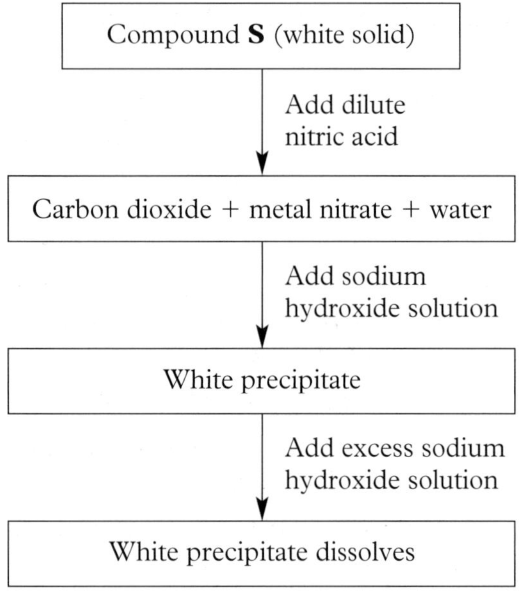

i) What type of chemical reaction occurs when nitric acid reacts with compound **S**? *(1 mark)*

ii) What does the reaction in (i) tell you about compound **S**? *(1 mark)*

iii) Which metal ion is present in compound **S**? Give a reason for your answer. *(2 marks)*

b) Reactions starting with iron(II) sulphate.

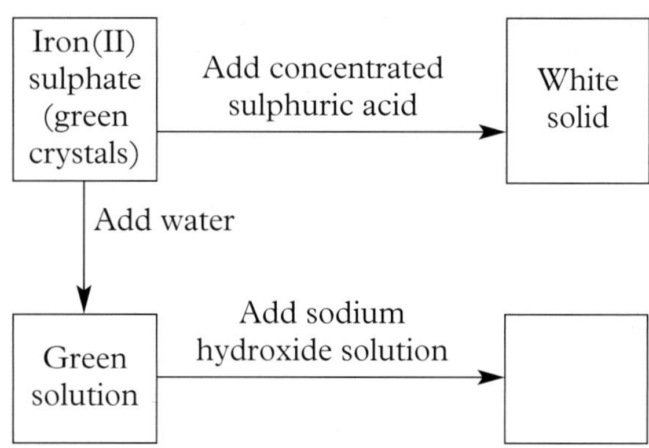

i) Copy the chart and write in the empty box the expected observation when sodium hydroxide solution is added to the green solution. *(1 mark)*

ii) Write a balanced ionic equation for the reaction in (i). *(2 marks)*

iii) What type of chemical reaction occurs when concentrated sulphuric acid is added to crystals or iron(II) sulphate? *(1 mark)*

2 Four bottles are known to contain the following substances of the same concentration.

sodium chloride solution ($NaCl$)
sodium hydroxide solution ($NaOH$)
sodium sulphate solution (Na_2SO_4)
ammonia solution (NH_3)

Unfortunately, the labels have come off the bottles.

Describe what **chemical** tests you would do to identify which bottle contained which substance.

Credit will be given for not only describing the tests and stating what you would expect to see, but also for the way you organise your answer. *(4 marks)*

Glossary

Acid A substance that dissolves in water to give a solution with a pH of less than 7.

It is a substance that forms hydrogen ions, $H^+(aq)$, when added to water.

Activation energy The minimum amount of energy needed by reactant particles before a reaction can occur.

Addition polymerisation A reaction in which unsaturated alkene molecules join to form saturated polymer molecules.

Addition reactions Reactions in which monomer molecules link together to produce a polymer and nothing else.

Air A mixture of gases made up of approximately 4/5 nitrogen and 1/5 oxygen.

Alcohol A member of a group of compounds containing an –OH group. Often used to mean ethanol, which is the alcohol found in alcoholic drinks.

Alkali A base (metal oxide or hydroxide) that dissolves in water to form a solution with a pH greater than 7.

It is a substance that forms hydroxide ions, $OH^-(aq)$, when added to water.

Alkali metals The name given to the metals in Group 1 of the Periodic Table.

Alkanes A family of hydrocarbons with the general formula C_nH_{2n+2}. Methane (CH_4) is the simplest alkane. Alkanes have a single covalent bond between the atoms.

Alkenes A family of unsaturated hydrocarbons with the general formula C_nH_{2n}. Ethene (C_2H_4) is the simplest alkene. Alkenes have a double covalent bond between two carbon atoms. Alkenes decolourise bromine water.

Alloys A mixture of metals (or of carbon with a metal). Alloys can have properties different from the parent metal(s).

Anaerobic A process which takes place in the absence of oxygen.

Anhydrous Crystals from which water has been removed.

Anions Atoms or groups of atoms that have gained electrons to become negatively-charged ions.

Anode The positively-charged electrode.

Anodising The process of making the protective oxide layer on the surface of aluminium thicker by making the object the anode in a bath of sulphuric acid.

Artificial fertilisers Fertilisers like ammonium nitrate that are not naturally occurring.

Atmosphere The layer of gases around the Earth.

Atom The smallest part of an element that can exist. Atoms have a nucleus consisting of protons and neutrons around which are shells of electrons.

Atomic number The number of protons present in an atomic nucleus (and the number of electrons present in the neutral atom).

Base An oxide or hydroxide of a metal.

Bauxite The main ore of aluminium containing aluminium oxide (Al_2O_3).

Biological catalyst A catalyst found in organisms – usually some type of enzyme.

Blast furnace The industrial method used for extracting metals such as iron, from their ores.

Boiling point The temperature at which a liquid turns to a gas.

Bond energy The energy (in kJ/mol) needed to break a bond.

Bonding The forces that hold atoms together.

Branched chain The arrangement of carbon atoms in organic molecules in which the carbon atoms are not all in a straight chain.

Bromine water A saturated solution of bromine dissolved in water. Used to distinguish between alkanes and alkenes.

Burette A narrow glass tube with a valve at the bottom to allow the dispensing of accurately measured volumes of liquid.

Carbon dioxide A gas formed during respiration and in the combustion of hydrocarbons. It turns clear limewater milky.

Carboxylic acids Organic acids containing the –COOH functional group.

Cast iron Iron straight from the blast furnace. Cast iron contains impurities including about 4% carbon.

Catalyst A substance that changes the speed of a reaction but remains unchanged after the reaction.

Cathode The negatively-charged electrode.

Cations Atoms or groups of atoms that have lost electrons to become positively-charged ions.

Charge A feature of atomic particles. Protons and electrons have a charge. Electrons have a negative charge and protons have a positive charge. Opposite charges attract; like charges repel.

Coke The substance used as the reducing agent in the blast furnace. It is almost pure carbon and is made by heating coal.

Combustion The burning of a substance in oxygen to release heat energy.

Complete combustion The burning process of organic fuels in which no carbon monoxide is produced. (The fuel is oxidised as much as possible.)

Compound A substance which contains two or more elements chemically joined together.

Contact process The industrial process which converts a mixture of sulpur dioxide and oxygen to sulphur trioxide. The process is used in the manufacture of sulphuric acid.

Core (Earth) The innermost part of the Earth.

Corrosion A reaction between a metal and substances in the atmosphere.

Corrosion resistant A term used to refer to steels that have a high chromium content (about 15–20%) and which do not rust easily. (Used to be called 'stainless steel').

Covalent bond The bonding of atoms caused by the sharing of pairs of electrons in their outer electron shells.

Covalent compounds Compounds in which the atoms are held together by covalent bonds.

Cracking A form of thermal decomposition in which large hydrocarbon molecules are broken down into smaller ones.

Cross-linking bonds Bonds between neighbouring long chain molecules in polymers.

Crude oil A mixture of substances, most of which are hydrocarbons, formed by the anaerobic decomposition of marine organisms over a long period of time.

Crust The surface layer of the Earth.

Cryolite Used with bauxite in the extraction of aluminium by electrolysis. It lowers the melting point of bauxite and increases the electrical conductivity of the bauxite.

Dehydrating agent Substances (like concentrated sulphuric acid) which can remove water from other substances.

Denaturing The process by which enzymes are destroyed when heated above a temperature of about 40 °C.

Denitifying bacteria Bacteria which convert nitrates in the soil into nitrogen gas.

Density A means of comparing the 'heaviness' of different substances. Usually quoted as the mass of a certain volume of the substance (g/ml or kg/m^3).

Deposition The laying down in water of a layer of rock fragments.

Displacement reaction A reaction in which one metal displaces another metal.

Double bonds Where two atoms are held together by two bonds instead of just one (e.g. C = C, C = O).

Electrolysis The process of splitting up a chemical compound using an electric current.

Electrolyte The solution in which electrolysis takes place.

Electrons Negatively-charged sub-atomic particles orbiting in shells around the atomic nucleus.

Electroplating The process of covering materials (usually metals) with a thin layer of another metal.

Element A substance made up of atoms which contain the same number of protons so contain only one type of atom, and which cannot be broken down into anything simpler by chemical means.

Endothermic reaction A reaction in which heat energy is transferred from the surroundings because more energy is needed to break the existing bonds in the reactants than is released when new bonds are made in the products.

Energy level diagrams Diagrams which show the energy content of the reactants and the products during a chemical reaction.

Enzyme A protein that can act as a catalyst for a reaction. It can be easily destroyed (denatured) by heating.

Equilibrium (reversible reactions) When the forward reaction proceeds at the same rate as the reverse reaction.

Esters Organic compounds formed by the reaction between a carboxylic acid and an alcohol. They usually have strong odours and tastes and are present as the flavouring in many foods.

Eutrophication A process caused when large amounts of nitrates and phosphates are discharged into rivers and streams. The nutrients cause the rapid growth of algae and water plants. The eventual death of the algae and plants soon leads to the rapid growth of aerobic bacteria. These decomposers soon use up all the available oxygen in the water. This in turn causes other animal life in the water to suffocate and die.

Evaporation The loss of the more energetic particles from the surface of a liquid.

Exothermic reaction A reaction in which heat energy is transferred to the surroundings because more energy is given out making the new chemical bonds in the products than is taken in to break the existing bonds in the reactants.

Fermentation The changing of glucose into ethanol (alcohol) and carbon dioxide by the action of enzymes in yeast.

Fertiliser A substance which can be natural or artificial applied to soil to improve the growth of plants.

Flame test A way of identifying the metal present in some compounds by means of the colour of a flame.

Flammable Easily set on fire.

Fluid A liquid or a gas.

Formula mass The mass in grams of one mole of a substance.

Fossil The remains or imprints of dead plants or animals trapped in sedimentary rocks when the rocks were formed. The remains or imprints may have been mineralised and turned into stone.

Fossil fuels The non-renewable energy resources: crude oil, natural gas and coal.

Fossilisation The process that produces fossils.

Fractional distillation A method of separating liquids whose boiling points are close together. The process used to separate the different substances in crude oil.

Free electron The electrons in metals that move around inside the metal and do not remain in orbit around a nucleus. The presence of these free electrons allows the metal to conduct electricity and heat.

Functional group Groups in organic molecules that have particular behaviours (e.g. –OH, –COOH).

Galvanising The process of covering iron or steel with a layer of zinc.

Gas chromatography A form of chromatography in which gases are identified by the speed they travel along a long tube.

General equation An equation in which a general formula is used not a specific one.

General formula A formula that represents a group of compounds (homologous series) (e.g. the general formula of alkanes which is C_nH_{2n+2}).

Giant structure Ionic compounds that have high melting points, usually dissolve in water and are good conductors of electricity when molten or in aqueous solution.

Global warming An international problem caused partly by the increase in the amounts of carbon dioxide and methane in the atmosphere which results in an increase in the average temperature of the Earth.

Greenhouse effect The effect in the atmosphere of heat energy being absorbed by gases such as carbon dioxide and methane.

Group A vertical column of elements in the Periodic Table having similar chemical properties due to the atoms of the elements having the same number of electrons in their outer shells.

Haematite A common ore of iron containing iron(III) oxide.

Haemoglobin The red pigment in the red blood cells which combines with and transports oxygen.

Halide The compound formed when a halogen reacts with another element.

Halogens The name given to the elements in Group 7 in the Periodic Table.

Hard water Water that contains calcium (Ca^{2+}) or magnesium (Mg^{2+}) ions. Hard water forms a 'scum' with soap.

Homologous series A group of organic compounds that have similar chemical properties (e.g. alcohols, carboxylic acids etc.).

Hydrated proton The name given to the hydrogen ion (H^+) when in water. It can be represented by the symbol $H^+(aq)$. (In some books it is represented by the formula ($H^+{}_3O$.).

Hydrocarbons Compounds containing only hydrogen and carbon.

Hydrogen The chemical element with the lowest density. Small amounts of it burn with a squeaky pop.

Hydrogenated A compound in which the C = C double bond has been reacted with hydrogen to form a C–C single bond.

Hydrogen ion An ion (H^+) present in all acids.

Hydroxide ion An ion (OH^-) present in all alkalis.

Igneous rocks Rocks formed by magma rising upwards from the mantle, cooling and solidifying into a hard crystalline rock.

Incomplete combustion The burning of organic compounds in an inadequate supply of oxygen so that carbon monoxide gas or elemental carbon are produced instead of carbon dioxide.

Indicator A dye which changes colour when mixed with acidic, alkaline or neutral solutions.

Inert Unreactive.

Insoluble A substance that will not dissolve in a liquid, usually water.

Intrusive rock Igneous rocks formed when lava cools beneath the surface of the Earth. Because the rock cools slowly, the crystals are large.

Ion An atom or group of atoms which have lost or gained electrons to become positively or negatively charged.

Ion exchange A process in which one type of ion is removed and replaced by another.

Ionic bond The electrostatic attraction between opposite charges responsible for holding metal and non-metal elements together in a compound. The ions are formed when the metal atoms transfer electrons to the non-metal atoms in order to achieve full outer electron shells.

Ionic compounds Compounds formed by the attraction between ions of opposite charge.

Ionic equation Equation which shows the ions taking part in a reaction. In ionic equations the charges must balance as well as the number of ions/atoms involved.

Ionise To remove electrons from or add electrons to atoms or groups of atoms so giving them positive or negative charges.

Isomers Organic substances with different structures but the same molecular formula.

Isotope Atoms of the same element which contain different numbers of neutrons.

Lava Magma that has erupted through the Earth's crust.

Law of Conservation of Mass This states that during any chemical reaction matter (material) is neither created nor destroyed.

Limescale The substance formed on the inside of vessels in which hard water has been heated.

Limewater A solution of calcium hydroxide that turns milky when carbon dioxide is passed through it.

Magma Molten rock below the Earth's crust.

Mantle The layer of the Earth between the crust and the core.

Mass number The total number of protons and neutrons in an atomic nucleus.

Mass spectrometry A method of identifying elements by measuring the mass of the atoms present.

Melting point The temperature at which a solid turns into a liquid.

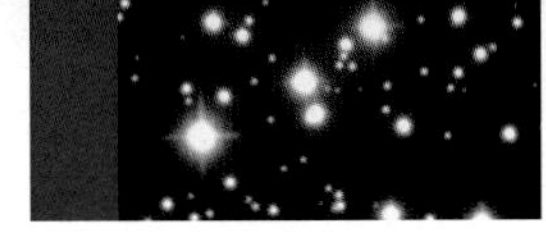

Metamorphic rocks Rocks formed from rocks which became buried deep underground and had their structure changed by high temperatures and/or high pressures.

Mild steel Steel that contains very little carbon (usually <0.3%).

Mineral A solid element or compound found naturally in the Earth's crust.

Mixture Two or more substances which are usually easy to separate.

Molar The word used to indicate the concentration of a solution (e.g. a solution with a concentration of 2 moles/dm^3 would be called a '2 Molar' (or 2M) solution).

Molar volume The volume occupied by 1 mole of any gas. The molar volume of any gas at room temperature and atmospheric pressure is 24 dm^3 (24 000 cm^3).

Mole The mass in grams of 6×10^{23} particles of any substance. It is the relative atomic mass of an element or the relative formula mass of a substance expressed in grams.

Molecular formula The basic formula of a compound (usually an organic compound) (e.g. C_2H_6 for ethane) that just indicates what elements are present and their quantity.

Molecule A particle containing atoms of the same or different elements bonded together. The smallest part of an element or compound that can take part in a chemical reaction.

Monomers Small molecules which join together to form a long chain of molecules called a polymer.

Negative The charge on an electron.

Neutral (charge) Having no overall charge.

Neutral (indicators) Having a pH of 7.

Neutralisation A reaction between an acid and a base or a carbonate.

Neutron A particle with no electrical charge found in the nucleus of most atoms. Its mass is similar to that of a proton.

Nitrates Chemicals containing NO_3 ions, frequently used in fertilisers to help plants synthesise proteins.

Nitrifying bacteria Bacteria which convert ammonium compounds in the soil into nitrates.

Noble gases The name given to the elements in Group 0 in the Periodic Table.

Non-metals Elements in the Periodic Table which usually have low melting points and boiling points, are poor conductors of electricity and heat, and as solids are brittle.

Non-renewable (finite) energy resources Energy resources that, once used, cannot be replaced.

Nuclear magnetic resonance A means of identifying the functional groups present in an organic molecule by the way the molecule absorbs energy when in a magnetic field.

Nucleon The protons and neutrons in the nucleus of an atom.

Nucleus The central part of an atom that contains positively-charged protons and uncharged neutrons.

Ores Minerals or mixtures of minerals from which a metal can be extracted in economically viable amounts.

Organic Compounds of carbon found in large quantities in living and dead organisms.

Oxidation A chemical reaction which involves the addition of oxygen.

A reaction involving the loss of electrons.

Oxygen The chemical element that is vital to life. It will relight a glowing spill.

Ozone layer The layer of gas in the upper atmosphere that reduces the amount of harmful ultraviolet radiation reaching the Earth's surface.

Period A horizontal row of elements in the Periodic Table.

Periodic table The arrangement of the elements in order of increasing atomic number.

Permanent hardness Hardness in water that cannot be removed by boiling the water. usually caused by substances like calcium sulphate.

pH A scale used to measure acidity and alkalinity.

pH scale A set of numbers from 1 to 14 used to measure the acidity or alkalinity of an aqueous solution.

Phosphates Chemicals containing PO_4 ions, frequently used as fertilisers to help plants photosynthesise and respire.

Photosynthesis The process in green plants which produces biomass (initially carbohydrates) and oxygen, and requires carbon dioxide and water as raw materials and chlorophyll to enable the plant to absorb light energy.

Pipette A calibrated tube than can deliver a very precise volume of liquid.

Plastics The common name for polymers.

Polymer A long chain molecule made up of many smaller molecules called monomers.

Polymerisation A reaction in which small molecules join together to make larger molecules.

Positive The charge on a proton.

Potassium An element used by plants to help the action of the enzymes involved in photosynthesis and respiration.

Precipitate The formation of an insoluble solid during the reaction between two solutions.

Precipitation The type of reaction in which a precipitate is formed.

Products The new materials produced as a result of a chemical reaction.

Proton A positively-charged particle found in the nucleus of an atom. It has a mass similar to that of a neutron and the number of protons present decides which element is present.

Proton acceptors Substances which can react with hydrogen ions (Usually bases.)

Proton donors Substances which contain hydrogen ions (Usually acids.)

Reactants The starting materials in a chemical reaction.

Reactivity series A list of metals arranged in order of their chemical reactivity. The most reactive metals are at the top of the list.

Redox reaction Reactions in which one reactant is REDuced and another is OXidised.

Reduction A chemical reaction which involves the loss of oxygen.

A reaction involving the addition of electrons.

Relative atomic mass The average mass of an atom of an element on a scale on which the mass of a hydrogen atom = 1 or the mass of the ^{12}C isotope of carbon = 12. It takes into account the relative abundance of different isotopes with different mass numbers.

Relative formula mass See relative molecular mass

Relative molecular mass (relative formula mass) This is found by adding together the relative atomic masses of all the atoms in one molecule of the substance.

Resources Natural materials available for the use of organisms.

Respiration The process taking place in living cells transferring energy from food molecules (glucose) to cellular energy.

Reversible reaction A reaction that can proceed in either direction depending on the reaction conditions. Reactants can be changed into products which in turn can be changed back into reactants.

Rusting The corrosion of iron in the presence of air and water to form hydrated iron oxides.

Rutile The name of a naturally-occurring ore of the metal titanium (mainly TiO_2).

Sacrificial protection Used to reduce the rusting of iron by attaching a more reactive metal such as magnesium or zinc.

Salt The name of any substance that contains a positive ion other than H^+ and a negative ion from an acid (e.g. 'common' salt, sodium chloride Na^+Cl^-).

Saturated hydrocarbons Hydrocarbons in which the carbon atoms are all linked together with single C — C bonds.

Saturated solution A solution that contains the maximum amount of solute that will dissolve at that temperature.

Scum The unpleasant 'greasy' material that forms when soap reacts with the calcium and magnesium ions in hard water.

Sedimentary rocks Rocks formed by deposition, burial and compression of weathered rock fragments or the shells of dead animals. They can also be formed by the precipitation of calcium carbonate usually in warm, shallow seas.

Slag The substance that forms in the blast furnace that contains most of the impurities.

Smelting The process of getting a metal from its ore by heating the ore with carbon.

Soap The salt of a long chain organic acid that is able to disperse oily materials in water.

Soft water Water with no (or very few) calcium and magnesium ions in it.

Solubility A measure of the amount of solute needed to produce a saturated solution. Usually measured as the mass (g) of solute that will dissolve in 100 g of solvent.

Solute The substance (often a solid) that dissolves in a solvent (usually a liquid) to produce a solution.

Solvent The substance (usually a liquid) into which a solute will dissolve to produce a solution.

Spectroscopy The process of identifying materials by the way they absorb or emit particular frequencies of electromagnetic radiation.

Steel An alloy of iron and carbon. (Usually containing up to about 1.5 to 2% carbon.) Specialist steels may contain other alloying elements like chromium or tungsten to impart specific properties.

Steroid A group of complex alchohols that include the sex hormone testosterone and the substance cholesterol.

Straight chain The arrangement of carbon atoms in organic molecules in which all the carbon atoms are in a straight chain with no branching.

Strong acid An acid which is almost 100% ionised.

Strong base A base which is almost 100% ionised.

Structural formula A way of displaying the formula of a compound so that the bonds and the approximate positions of the atoms are shown. Usually only used for organic compounds.

Synthesis The process in which elements are chemically combined to make a new compound.

Temperature How hot or how cold an object is. Units are °C.

Temporary hardness Water hardness that can be removed by boiling the water. Usually caused by calcium or magnesium hydrogen carbonate.

Thermal decomposition The breaking down of a compound by the action of heat.

Thermit process A method of joining two lengths of railway track together using the exothermic reaction between aluminium and iron(III) oxide.

Thermosetting plastics Polymers which have to be heated to make them cure (set).

Thermosoftening plastics Polymers that soften when heated and can be moulded into shape.

Titration A method of determining the concentration of a solution by neutralising it with another solution of known concentration.

Transition elements (transition metals) The name given to the elements in the Periodic Table between Groups 2 and 3.

Transition metal See transition elements.

Transportation The removal of rocks broken down by weathering and erosion.

Universal indicator An indicator used to measure the pH of a solution to show whether the solution is acidic, neutral or alkaline.

Unsaturated hydrocarbons Hydrocarbons in which some of the carbon atoms are all linked together with C=C double bonds.

Water cycle The process by which water on the Earth circulates. Water in the sea evaporates into the air. It cools and falls back as precipitation (e.g. rain).

Water of crystalisation The water contained within the crystals of some substances.

Weak acid An acid which is almost totally unionised.

Weak base A base which is almost totally unionised.

Weathering The chemical, physical or biological action by which rocks are broken down into rock fragments.

Wrought iron Almost pure iron.

Yeast A naturally-occurring micro-organism that produces enzymes that cause fermentation.

Index

Note: Glossary entries are in bold